ECO-FASCISTS

ECO-FASCISTS

How Radical Conservationists Are Destroying Our Natural Heritage

ELIZABETH NICKSON

BROADSIDE BOOKS
An Imprint of HarperCollins*Publishers*
www.broadsidebooks.net

HarperCollins books may be purchased for educational, business, or sales promotional use. For information, please write: Special Markets Department, HarperCollins Publishers, 10 East 53rd Street, New York, NY 10022.

FIRST EDITION

Library of Congress Cataloging-in-Publication Data has been applied for.

ISBN: 978-0-06-208003-5

12 13 14 15 16 OV/RRD 10 9 8 7 6 5 4 3 2 1

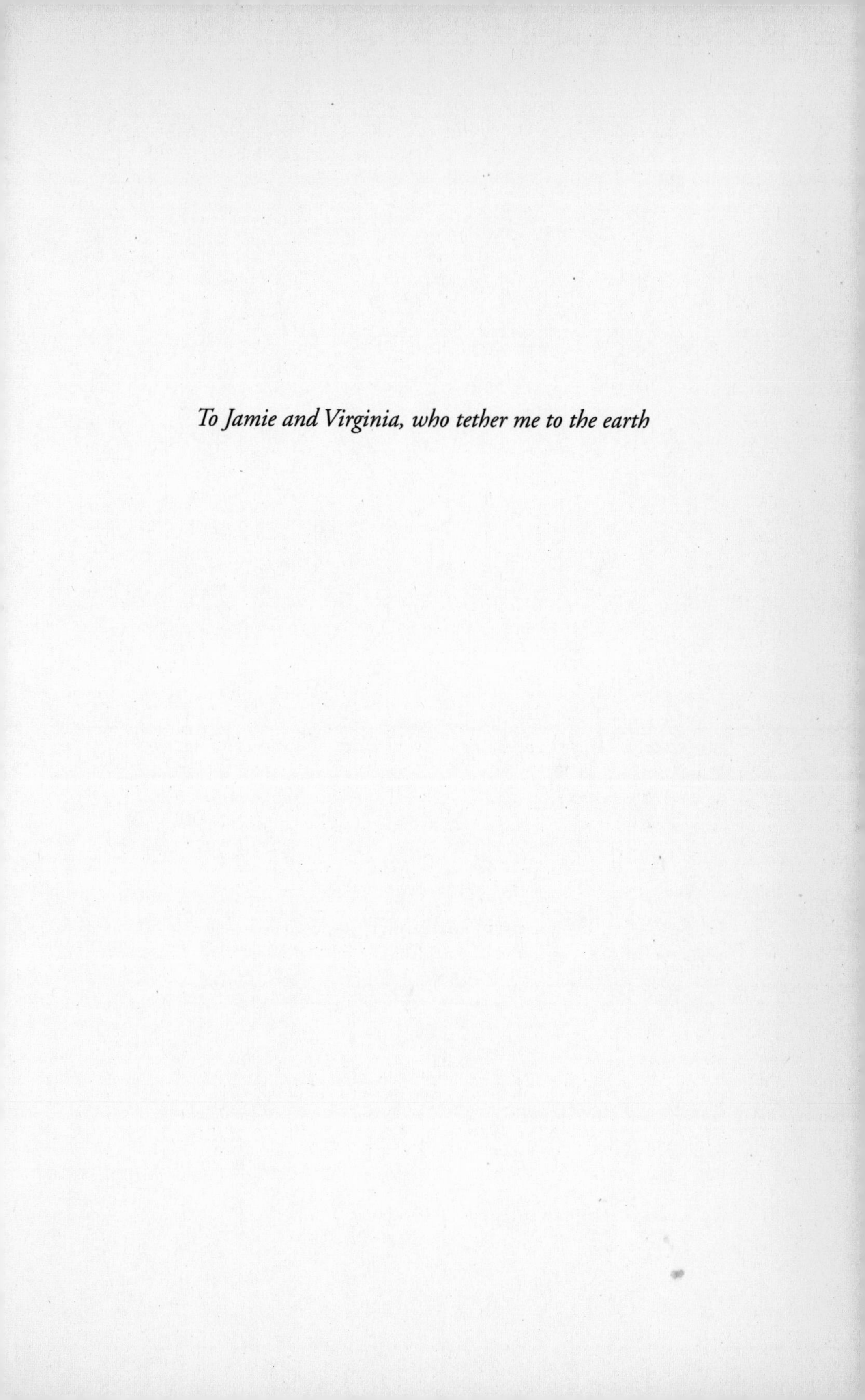

To Jamie and Virginia, who tether me to the earth

CONTENTS

PREFACE

Most criticism of the environmental movement, well founded or not, emanates like a cloud from economists and rationalists at think tanks who work principally with data and statistics. I am neither of those things, though I am very fond of numbers and reason.

By contrast, insofar as possible, I lived this book. In fact, I walk the green walk more than anyone I've met. I live on sixteen acres of field and forest and creek. More than half of my land—an older-growth forest and magnificent ravine—is locked down in a perpetual "no-touch" covenant. I have a salmon enhancement project at the intersection of my two creeks. I am restoring two meadows. One used to be a gravel pit; the other, when I began work, was entirely covered by impenetrable Scotch broom, six feet high. Broom is an aggressive invasive shrub that has destroyed more indigenous plant life in my bioregion than any development, and it's all I need for a gym for the rest of my life, because broom apparently never really dies.

I live in a green house heated by geothermal, and with the exception of the propane I use to cook, the house is carbon neutral, even though I think climate science is so corrupted that we have to go back to the beginning and start again. The house is structural rammed earth, which means it's built out of dirt from the gravel pit a mile down the road, though the architecture owes more to Palladio than Middle Earth and I am a girl, so it's a deep rose color. The house has enormous thermal mass, so it costs 50 percent less to heat than a conventional house of its size, and during winter power outages, it takes five days to lose its warmth. We have green roofs and an on-demand hot-water heater, among other green enhancements. The house has no drywall, no paints, no solvents, and could be certified as a healthy house.

Clearly I have integrated many of the dictates of the modern environment movement into my daily life. From this stance, from an open-hearted conformity to what I believe is a grand idea corrupted by powerful fanatics, I was able to understand the movement's many disastrous failings.

I am also from an old country family, one of the founding families of the American republic, Puritan pioneers who arrived in 1630 "praying and expounding the word of God the whole way,"[1] to create what they hoped would become Christ's millennium on earth. Deliberately describing themselves as late as the nineteenth century as being "of the people,"[2] the Phelps clan "hung on their cooking pot, cooked their homely meals and went down to the founding of Windsor" in the Connecticut Valley in the 1650s. From there they spread, pioneering in every corner of the continent, bringing with them a hearty appetite for hard work and a hands-on local charity unrivaled today even by the standards of Mr. Gates; it was no small matter to be a businessman, community leader, *and* an officer of the Underground Railroad in the 1830s, '40s, and '50s, hiding fugitives in dugout cellars and tunnels under your house. Ideas inspired them; dozens fought in the revolutionary army, one was the spy at Fort Ticonderoga. Another, the initial county and township planner of the new republic in the 1780s, was deputy commissary of the revolutionary army. That coding lies in millions of us in the Americas, and it birthed the finest ideas the human race has ever developed—ideas that created staggering bounty and the greatest egalitarian nation the world has ever seen, ideas that have been replaced with an uncompromising and rigid bureaucratic command-and-control structure, which is creating yet another hybrid of the totalitarian state.

To say I was uninterested in my heritage before I began this book would be an overstatement, but as I grappled with the Leviathan we have created and the attendant research, I found myself sent back to their diaries and history again and again, for relief at first, then because the ideas were clear, beckoning to freedom, self-determination, and a comity that I frankly envied. Their ideas made sense; in contrast, what I was attempting to understand clutched at me like a wet, dark, extinguishing cloud—so restrictive, punishing, and narrow that some days I was genuinely frightened. Some days, too, I felt my family reaching right through me, furious and despairing at what we are making. The principles that drove them through difficult and terrifying times began to propel me. As I peeled back each layer, I found myself growing implacably angry and opposed in a way I had never experienced.

I was trained as a reporter on the job at *Time* magazine, and in my seven years at the London bureau, Time Inc. sent me into disparate worlds, one after another. I went from deposed royalty to torture victims to grandees of every sort to Nelson Mandela's release, where I helped the company acquire his memoir. I learned, sink or swim, to get my bearings very quickly wherever I was and bring home the story. Reporters like me are notoriously hard to domesticate because they are habituated to the adrenaline of the chase, so the mounting horror I felt in every succeeding month was familiar stuff. Instead of spitting the lure out, I became further engaged.

The environmental movement has created a bewildering array of institutions that have been well hidden behind the scrim of modern life. Private foundations, the most pivotal in American culture, have been largely captured by a subset of grim zealots seeking to remake the world. They are joined by fervent true believers in federal and state agencies like the Department of the Interior, the Forest Service, the Environmental Protection Agency (EPA), the Fish and Wildlife Service, and a proliferation of others. These bureaucrats, if they miss a beat, are spurred, litigated against, and generally hagridden by nongovernmental organizations (NGOs), which marshal and mobilize tens of billions of dollars every year in the United States alone to promote their agenda. The complexity they have created is almost impossible to grasp in its entirety; it is a suffocating web of lies, distortions, fearmongering, and bad science married to top-of-the-line strategic planning, which has triggered an error cascade[3] wreaking destruction everywhere prosecuted.

So what were the results, I asked myself, of this new green world? Were there any tough audits of the results? How were the ranges and forests faring under green command? What about the people? The next step became obvious. I jumped into my car and set out to meet the men and women who grapple with the beast daily, because otherwise their families would not eat. It was astounding to me that only one journalist in all of the United States, Timothy D. Findley, who died last year, had done systematic on-the-ground reporting of this story. While I didn't take his routes, I drove in his tracks, experiencing his bleary-eyed exhaustion as I pulled into the parking lot of a Best Western in a town of 3,452 somewhere on the Colorado–New Mexico border at one in the morning.

I drove deep into America's backcountry, through blinding snowstorms, over mountain passes, across black ice, and through torrential rains, all the way through Utah, Idaho, Wyoming, and Colorado twice, down to the Mexican border in California, and through New Mexico's boot heel to Texas. While those twelve or fifteen thousand miles should have been lonely and sometimes frightening, they were neither, not even along the most dangerous parts of the Mexican border. The country behind the strip malls and city access highways is so magnificent and the people so kind, funny, and warm, it was like driving through a lost paradise filled with newly met friends. If I were thirty, I'd buy a string of cows and horses and vanish into Wyoming or Nevada or Idaho, never to be seen again.

While part of the book can be described as a personal journey, it is so only in that I experienced, in my own grappling with the green gods, just about everything that is being forced on every rural inhabitant of the world. As absurd as this might sound, in its essentials what I experienced is duplicated by the struggle of a Thai forest dweller or a sheep farmer in Sussex or anyone anywhere who scrimped over a lifetime to buy oceanfront property and is now realizing his dream is dead and his money, gone. Multiply my experience by 100 million and you will understand what happened and how. The incidents in that part of my journey come from a plan, one that has been carefully devised and put in place over the past thirty years. Those planners have created an entirely new culture in rural areas, a culture of deliberate decline that has not only already damaged the suburbs but plans to eradicate them. Suburbanites will, over the next decades, be methodically moved into the cities, which will become choked, toxic, and controlled by an iron system of regulation being implemented right now. Therefore the book is structured to illustrate each step of the process by which the property rights and agency of that sheep farmer, forest dweller, or middle-class writer are removed. Interstitial chapters that describe an aspect of my own experience are followed by a chapter in which the larger implications of that experience are examined, looking at history, legislation, science, and ultimate results in both human lives and the natural world, including the calamitous effects on the larger economy.

I know this story because I've lived it every step of the way. By the end of the book, you will know it too.

The title of this book is harsh, particularly when used with regard to environmentalists, whom most people view as virtuous at best, foolish at worst. But I do not use this term lightly, nor as a banner to grab attention. My father landed on D-day and, at the end of the war, was put in charge of a Nazi camp and told to "sort those people out." He was a tender soul despite the heroism, and my brothers and I grew up intimate with his nightmares—nightmares with him until he died. That darkness and history taught me that man defaults to tyranny over and over again, and while the tyranny of the environmental movement in rural America has not reached what its own policy documents say is its ultimate goal—radical population reduction—we cannot any longer ignore that goal and its implications.

Nor can we ignore the exigencies of creating the grail of "a sustainable America," for they are staring us in the face: coercion, regulation so punitive entire counties have been bankrupted, property confiscation, the loss of ancestral homes, the separation of people from loved lands and professions, and, above all, financial catastrophe that has been visited on rural family after rural family from sea to shining sea. And the full program of sustainability, its metrics designed in swarms of conferences in every profession and sector, is being advanced in the suburbs and cities by a large, enthusiastic, and reckless army.

Like all bad ideas, sustainability is based on bad science. Conservation biology, a new iteration of biology, was brought into being at the University of Southern California in the 1970s and has enjoyed a long and powerful reign, coming to steer land use almost everywhere. Not long enough to be tested, apparently, because it has received very little testing, but long enough to be audited. The results are disastrous. Conserved ranges are desertifying, conserved forests dying, and watersheds being so badly managed that that magnificent triumph of civil engineering—the dams, waterworks, and irrigation of the American river system—is being overwhelmed. And, from the Serengeti to New Mexico's boot heel, wherever people have been cleared from the land, biodiversity collapses. It is axiomatic.

The sustainability army, while made up of seemingly ordinary, well-meaning middle-class people, is fast becoming the enemy within. Believing we are destroying the planet, it acts on that imperative and, as a result, lacks moral imagination; it does not see the harm trailing in its wake, and if it does see the human harm, thinks that harm necessary. As a result, the sustainability army is implacable; it hates traditional American values, believing those values are evil. It seeks to supplant them, and in some cases, destroy any evidence of American culture. I lost count of the number of times rural men and women in every corner of the country told me they hoped they would die before they had to witness rural people being killed. They are just that frightened. Crashing the founding culture of a country—which almost always is a culture bound to the land—is nation busting. While Jefferson's nation of yeoman farmers has been long superseded by industry and technology, it didn't go away. Out of sight, it remained the bedrock of our freedom, and it is being destroyed.

CHAPTER ONE

We Have Been Fooled

The people wish to be deceived, so let them be deceived.
(Mundus vult decipi, ergo decipiatur.)
—ANONYMOUS, ANCIENT ROMAN

No wonder the epigraph above was Hitler's favorite proverb, for by the time people woke up, the world was at war. Today we are nearing an inflection point of a different order. America's rural people are locked into a twenty-five-year cold war for survival. Their enemy? Us. Or rather an idea, held by us, both foolish and near universal, which has nearly destroyed their lives and is bringing ruin to the forests, fields, and watercourses upon which we all depend. Today, that rough beast is slouching toward the suburbs and cities, tyranny—soft tyranny but tyranny nonetheless—in store for us all.

What are the signs? At the end of 2011, extreme poverty spiked higher than any time since the Great Depression. This poverty is disproportionately rural. Cataclysmic floods and fires sweep the heartland almost every summer, gutting wealth. Forty million people have been driven from their traditional rural occupations, their final destination, if the movement has its way, into "pack 'em and stack 'em" housing around big-box stores and strip malls on the outskirts of cities. The western forest, the most productive forest in the world, is dying. America's rangeland is dying. The water system, that great engineering feat of the early twentieth century, is being deliberately broken. Country people believe that Russian cadaver wolves, weighing almost three hundred pounds, trained to clean up battlefields of the dead and dying, have been introduced into conserved forests and wilderness. Confirmed wolf kills of humans are on the rise.[1] Stray humans, family pets, and livestock are far easier to bring down than elk and antelope, which are disappearing anyway, because the forests are overgrown and open meadows vanishing.

Rural America is riven by meth, pretty much the only growth industry, aside from, of course, debt. Government jobs—principally in prisons and elder care—are on the increase, but they only manage to keep rural America on life support. On Wall Street, credit default swaps are necessary because they insure the credit all of us in the hinterlands need, because a predictable middle-class income—for those not included in protected environs of university, media, government, and the financial sector—has become an ancient dream.

While not the entire reason, by any means, what I write about is the hidden factor, the festering sore from which the putrefaction grew, and that is our culturewide delusion about the collapse of our natural systems, the loss of land to industrial predation, and the crash of biodiversity caused by humans and development. All this cultural destruction is necessary, we tell ourselves—when we notice it at all—because if we took off the brakes, we would eat the world alive, and our grandchildren would scrape a living on a blasted desert.

This delusion has led to a sequestration of productive land unmatched since the age of kings. Over 30 percent of the American land base lies under no-use or limited-use restrictions, almost 700 million acres. The Bureau of Land Management and the Department of the Interior are targeting for confiscation another

213 million acres, bringing the count to nearly half the continent locked down.[2] In the developing world, we—and it is *we*, not the citizens of those countries—have sequestered an area greater than the continent of Africa. All those lands hold valuable natural resources, which could lift indigenous families out of desperation but which have been shut down, by the best and brightest among us, in perpetuity.

Commodity prices have risen almost 600 percent since 2000, hurting all of us, but principally the least advantaged among us, the urban, rural, and suburban poor.[3] While financialization is the accepted reason, I argue that the domination of leverage is made necessary by the gutting of the resource economy. There is no faster way to jack up prices than to create artificial shortages.

On lands remaining in productive use in the developed world, we have proliferated regulation so complex and absurd that an applicant without patronage can enter the maze and stay captured, bleeding his family's wealth, for decades. In 2000, Peruvian economist Hernando de Soto and his team traveled the developing world to identify the barriers to growth and found that the missing link was property rights.[4] For a man living on a scrap of dirt or trading out of a tin hut in a barrio to achieve title to that scrap or hut—which would make possible small loans, and therefore growth—would take, on average, eighty years and cost hundreds of thousands of Yankee dollars.

We are approaching that unhappy state in the developed world, the result of which has stunted both individual and collective wealth. There is no faster way to create a nation of serfs than to remove property rights.* Nor can you confiscate a country's assets without damaging the income of every single individual in that country, excepting, of course, that of the confiscators.

* Little wonder, then, that in concert with our metastatic conservation, capital investment in America has declined for twenty years and has now entered negative territory. The capital stock of the country is being depleted.

• • •

How did this happen? The answer is, of course, money.

The most august foundations of American capitalism, the Pew, Rockefeller, Heinz, Hewlett, and Moore foundations, along with the incubator, the mysterious Tides Foundation, the Nature Conservancy, and scores of others, spend more than $9 billion a year[5] coordinating the effort to turn rural people off their lands and curtail the lives of city and suburban dwellers. That $9 billion is spent selling fear. Its power is almost total; in fact, as market researcher Alan Dean points out, Microsoft spent $400 million introducing Windows 7, one of the biggest rollouts in the last ten years, but the environmental movement spends that *every single month.** Rather than virtuous David, in the marketplace of ideas the movement plays Goliath. The drumbeat of doom echoes from every conceivable medium. Little wonder we are all scared out of our pants.

It is impossible to fully understand this story without taking a full snoutful of the stench of class. Savor it, for you will be smelling it a lot; think of it as truffle oil, rich, stinking of decay, and very expensive. And because it is the twenty-first century, we're not talking robber barons. We're talking Tom Cruise and Teresa Heinz, Bob Redford and Ted Turner, our nouvelle oppressors luxuriating at the top of the world. One has only to watch the sneaking vanity of Tom Cruise going all Uriah Heep in front of his thousand-mile view at the top of his Colorado mountain, where he and his family gather to roast hot dogs and marshmallows, to see what is currently viewed as the apex of achievement.

* In 2008, the Urban Institute teamed up with the Rockefeller-founded Environmental Grantmakers Association to study the finances of the environmental movement and found that in 2003 the top fifty spent almost $5 billion a year. The average corporation spends between 1 percent and 3 percent of gross revenues on marketing and sales, but as Alan Dean points out, all any environmental nongovernmental organization does is sell its message. Five billion dollars divided by twelve months means that the environmental movement's marketing heft outclasses that of any major multinational.

In most environs, conservation has replaced breeding, money, and education as the social marker that distinguishes the individual from the herd. Saving land is considered an unalloyed good. As the drumbeat deafens, emotive language is attached to the stunning photographs of natural beauty that wallpaper our world. Little wonder that it is imperative in any of the big games of the superculture to have "saved" thousands of acres from industrial predation.

If only it weren't the industrialists doing it. If you take away anything, take away this: the dominant environmental aggressor in Canada's oil sands is the Pew Foundation. But Pew's grandfather, its original funder and creator, Sun Oil, now split off and named Suncor, is a prime extractor in the oil sands. This is not merely irony, say critics on the environmental left, pointing out that seven of the twelve board members of the Pew Charitable Trusts are either family members and heirs to the Sun Oil fortune or a former CEO. While no one is charging coordination between Suncor and Pew, Pew's activities increase regulatory costs in the oil sands, thereby shutting out smaller operators.[6]

Likewise, last year's Food Safety Modernization Act ensured that ADM (Archer Daniels Midland) and Monsanto improved their market share because family farmers had just been forced to layer thousands of dollars of regulatory costs on top of already fragile income statements. Little wonder Monsanto lobbied *for* the law.

The cumulative result is a catastrophic loss of rural population everywhere, people who, while undereducated by the standards of the cognitive elites, have traditionally stewarded their homelands well. In psychological terms, we have all become alienated from the land, moving in just a generation from a deep love of the land, ownership, pride, and joy, which was the norm for almost all of us as late as the 1960s, to despair. Instead of that heartening affirmation, we have turned on ourselves and see ourselves as original sin, a stain, a virus on the land. Best that we tread lightly, if at all. Preferably not at all. Only then can the land recover.

Entirely wrong. Man needs nature, but just as basally, as fundamentally, nature needs man.

So finally this is a story about our reckless meritocracy, which has depended, for decisions affecting the health of the land, on desktop modeling. Despite the obvious failure of the Wall Street algo traders, the perversion of climate-change science,[7] the cataclysmic failure of debt derivatives—all based on models—we still think that equations will save us. But here's the thing: nature is always and ever changing and owes more to chaos theory than any crystalline perfect state. In the natural sciences, a corrupt idea—that an ecosystem has to be in balance, with all of its members present in the correct proportion, to be considered healthy—has destroyed sensible stewardship of the natural world. There is no such thing as freezing an ecosystem with all its members present and in perfect balance to create health. It cannot be done, because as I will show you, so-defined healthy ecosystems do not exist, except in the lab or on the desktop. Yet the goal of creating a series of perfect ecosystems is reason and purpose for the vast reconfiguration of land use we are experiencing.

Bad science is immeasurably destructive. And conservation biology, because it is advocacy science that starts with the assumption that we are running out of everything, is bad science. It does not start with a clean slate, a perfectly neutral mind. It starts with an assumption, a reasonable assumption but an assumption nonetheless. I advance an alternate, evidence-based theory that abundance is the natural state of the world and that, for the most part, under democratic capitalism, man has created more abundance in geometric increase with every generation—until now.

In 2008, forester and professor Holly Lippke Fretwell spent a year trolling through the Forest Service's audit of the health of 446 million acres under its and the Bureau of Land Management's command. She discovered—and this was hard evidence, *not* modeled—that the forests conserved in service of the spotted owl are dead or dying. Almost 90 percent of the most productive forests in the world have been shuttered, and without any mitigation or maintenance, the closed forests were overstocked and pest ridden. Five hundred weakened, spindly trees grow where sixty to eighty healthy ones used to flourish. Two hundred million acres of once healthy forests is about to explode in a once-in-a-millennium conflagration that will extirpate species and desertify the forests. Not even the dirt will survive the forest apocalypse that beckons, much less seed stock, fish, wildlife, and plant life.[8]

Similarly, rancher and range scientist Allan Savory found in 2011 that millions of acres of preserved grasslands in the heartland are turning to desert. Grasslands need cloven-hoofed animals to thrive; their hooves break up the soil, bringing nitrogen and oxygen to grass and forbs. In addition, those once productive rangelands all had windmills, which pumped water not only to ranch animals but also wildlife for miles around. Those have been shut down; the wildlife has fled. Wherever you look, on saved lands, in both the developed and developing world, biodiversity is crashing.

By 1970, engineer Chuck Leaf had been working in the U.S. Forest Service for ten years; his superiors told him he was marked for rapid promotion. A bench scientist, he had a project that promised to increase water quantity to downstream communities in Wyoming and Colorado, an idea that could then be tracked out to include all western watersheds. "I started being sent to meetings in D.C., where we were told that we were to get as much land out of the hands of private owners as possible," he said. Then the service shut down his lab and started asking him to positively peer review science coming out of the universities. "It was crap, junk science," says Leaf today. " 'We have big plans for you, Chuck,' I was told, 'but you're going to have to get on board.' I left." Today on the Colorado front range, Chuck Leaf has to replace every cup of water he draws from his well, because of "water shortages."

Hydroelectric dams, which produce cheap, clean, renewable energy, are being methodically blown up, 925 as of the beginning of 2012, releasing tons of toxic sludge, devastating surrounding farms and ranches, destroying energy production, and draining entire agricultural regions of water. "Water is not a right, it's a privilege," announced the president in January 2011, and water rights going back to 1866 are now being revoked, despite the fact that rural people say that confiscating water in the country is tantamount to genocide.

Shortages, fires, desertification, and floods caused by the environmental movement and science perverted by bad politics—these are the true legacy of Earth Day.

Nor are exurban and suburban neighborhoods exempt. The weapons were honed and polished in rural areas, but by the late 1990s, an army of bureaucrats, foundation employees, and environmental activists advanced on suburban America. Their goal? Greening the suburbs. Whether it is called sustainable development, visioning, smart growth, or the UN's International Council for Local Environmental Initiatives (ICLEI, pronounced ICK-lee), suburban America is being collapsed from within. In fact, demographer and land-use expert Wendell Cox found toxic mortgages were concentrated where there was "sustainable" land-use regulation. Cox and his associates, after analyzing Federal Reserve and Census Bureau data, believe that instead of a $5 trillion housing bubble, without "sustainable development" regulation, a more likely scenario would have been, at most, a $0.5 trillion housing bubble.

Sustainable is the catchword of every activist, conservation bureaucrat, and green NGO officer. No one really understands what it means. Here's a clue. The following are considered unsustainable by the sustainability army: single-family homes, paved roads, ski runs, golf courses, dams, fences, paddocks, pastures, plowing of land, grazing of livestock, fish ponds, fisheries, scuba diving, sewers, drain systems, pipelines, fertilizer, and wall and floor tiles. These and many other elements of life as it is now lived are on any sustainability general's list for eventual elimination.

Little wonder that rural people warn us that our lives are being reconfigured without our permission or even notice. This despite the fact that there is no instructive example in history in which sequestering lands in the hands of the privileged few—which is, propaganda aside, exactly what's happened—has created anything but decline. The wealth of the Americas was founded on the right of the ordinary free man and woman to apply labor to property and create wealth. Nor is there any instructive example in history in which collective ownership—state ownership—of land has created anything but decline—decline in the wealth of the people, decline in the bounty of the land, decline in the health of the land, decline that is entirely unnecessary.

It is long past time, say America's rural people, for polite society, today's good Germans, to wake from our culturewide dream.

CHAPTER TWO

Green Fantasy Island

Where the bee sucks, there suck I. In a cowslips bell I lie.
—SHAKESPEARE, *The Tempest*

I came to this story late and by accident. While working in London, instead of a flat or condo, I bought with a friend a thirty-acre forest a half-hour ferry ride from my parents' retirement home, on an island in the Pacific Northwest.

I'd grown up in the country, in southern Quebec, twenty miles from the Vermont border, so by the time I came home, I was hungering for country life as I had for nothing else but children and the ability to write a decent sentence. I'd lived in Toronto, New York, and London for twenty years as a full-on urbanist. But one day, ten years after I'd bought the land, I found myself heading for a traffic island outside the Time Inc. building on Connecticut Avenue in Washington, D.C. Even though it was November, I kicked off my heels and planted my stockinged feet on the wizened grass, feeling for just that moment less parched, less metallic. Traffic roared by, but what I was doing was so odd, so out of time, that no one seemed to notice me. A year later, as if that moment had been a rehearsal, I vanished from the grisaille world of painful shoes and persistent anxiety and found myself ankle-deep in grace.

The ostensible reason was that my father was dying and I moved home to help. But that summer, after I built a cottage at the top of my hill, I found myself helplessly in love with the island and resolved to stay. After we had clear-cut an acre, I found that I had the proverbial hundred-mile view, from the coastal range that brackets Vancouver to the anime cartoon that is Mount Baker near Seattle, Washington, slices of navy blue sea between.

The island, Salt Spring, was on the large side—seventy-four square miles—and had about ten thousand residents, a figure that ballooned to twenty thousand every summer. There were three villages, and in the center of the largest, a lively and rightly famous market was staged every Saturday. The island had been identified by an overenthusiastic journalist as one of the top ten art towns in North America. Its legendary beauty had drawn people from the United Kingdom, France, Holland, Germany, Australia, South Africa, and every state in the Union, making for an interesting cultural life; many were here because regional politics are notoriously wacky. Postmaster General James Farley joked in 1936 that Washington state should be called the Soviet of Washington; that would go double for us. Besides, living was easy; even addicts and the homeless migrated here from all over the continent for the weather and the famous handouts.

The island was also known for its ecofeminist witches, who celebrated the high holidays of Wicca in their circles, casting petals on the water during the full moon and calling pods of killer whales (it was said) with their chanting. So while remote by the standards of London or New York, Salt Spring was not without interest for a former sophisticate.

In my first month, my loan officer had introduced me to Emma, a sardonic Londoner who, at eighteen, had married a fellow theater-school student. Her new husband became a megafamous rock star shortly thereafter. Two children and a few years later, she was divorced and rich for life. Emma emigrated to the West Coast to raise those kids and then went on to Salt Spring, where she built herself—at the top of a forested mountain, on two hundred acres near a thousand-acre park—a house straight out of an Andrew Jackson Downing pattern book. It was, as Downing—the Martha Stewart of 1850—advised, vast and comfortable but embellished by all the cozy cottagey signifiers that nineteenth-century suburbanites loved.

Then she found herself some "work."

Without the wives, daughters, and to a lesser extent, sons of the rich, the environmental movement would not be anywhere near the titanic force it is today. It is almost axiomatic that, just as Hare Krishna or Sri Chinmoy cults used to scoop up every shucked and miserable rock wife in the 1960s and '70s, on the courthouse steps today wait the Sierra Club, Greenpeace, and the Natural Resources Defense Council. What do they offer? A social life, for one thing, and a cause that will ease the pain of being thrown out of paradise. In fact, for the movement, nothing is more valuable than a woman with two, ten, or a hundred million dollars and animus to express. Just south of me, for instance, the heiresses to the KING 5 TV syndication fortune redesigned their parents' Bullitt Foundation and created a fierce environmental advocacy outfit, run by Denis Hayes, a cofounder of Earth Day. The original fortune was built logging the productive forests of the Pacific Northwest, but for years Hayes has funded nearly every activist group in nearly every forested community west of the Mississippi, the average grant of $10,000 in the hands of activists and environmental nongovernmental organizations (ENGOs) acting like a battering ram slamming into the economy of one county after another. But I am getting ahead of myself.

The night we met, Emma was, as Emma so often turned out to be, sullen. She scowled at me across the café table, fiddled with the menu, and declared she wasn't hungry—her subtext being that I bored her, and "who the hell was I anyway?" I floundered; I needed friends, and besides, God help me, I found the sulking charming. Looking for something to please her, I finally hit upon a meeting taking place that evening that she'd mentioned earlier. Why didn't we go together? That earned me my first smile. Off we went to Fulford Hall, a giant feed barn long ago converted to a community hall.

We were late and crept in, shoes in hand. Immediately Emma squeezed herself into a line of chairs along the back wall and sat down between a large bearded man and a hippie chick without another glance at me. I crossed the hall, found a spot, and tried not to feel abandoned.

My hurt feelings lasted about thirty seconds as the show began to unfold. Apparently the eighties disco star Princess Gloria von Thurn und Taxis, upon the death of her elderly cousin/husband, had begun to divest herself of her vast tracts of forests in the New World. One of these forests was on Salt Spring. She needed to pay death taxes, so two thousand acres on the south end of the island, which had been managed for decades using old-fashioned, ultrasensitive German silviculture, had been sold to a logging company with a slash-and-burn reputation. Several hundred furious residents were here to stop any such thing.

War drums sounded under the rustles and snorts and chair scrapings of the audience. I tried running a count, but my concentration kept breaking on the faces and their expressions. Sixty-year-old men with long stringy ponytails sat interspersed with the kind of country folk I grew up among—strong faces and much-worn wool and corduroy. A clutch of once gorgeous folk dresses from every conceivable indigenous culture, worn by women who, while older, were still stunning. Hippie mothers with children spilling out of their laps. Handsome disheveled men in their twenties and thirties wearing drapery not dissimilar to that of the Taliban. Distinct smells to puzzle out: wood smoke, patchouli, jasmine, pot, sweat.

Up at the front, a large old-fashioned pull-down screen showed a map of the logging company's holdings on the island. On a smaller illuminated screen, the company's mission statement scrolled by: "sensitive," "sustainable," "patch cutting not clear cutting"—all the approved phrases. Three men stood up there, neatly dressed and brushed and washed, their expressions cycling between grim determination and smiling, unctuous flexibility. They, too, they hinted as they strutted about, working their equipment, pointing out certain areas they were going to "save" and "preserve," had an iron fist. But they were jumpy, eager to please, and it seemed to me the crowd smelled that and were gleeful as they ran the developers through what must have been a carefully devised gauntlet.

The apparent sticking point was that the company was bent on cutting the dominant landscape on the island, ruining a view that everyone thought of as his own. The Fulford Valley was a configuration of mountain, rock face, fields, and forest that so perfectly represented a pastoral ideal that it defined the island for its visitors as they drove off the ferry. Not only did the company plan to shear giant patches of the forest on the mountains surrounding the valley, but on those patches, it planned to build post-and-beam trophy homes. I sank to my heels on the floor and watched, absorbed, as person after person stood up to make surprisingly sophisticated arguments against this scheme, managing to rebut or question every point the interlopers offered. Science mixed deliciously with emotion kept the developers off guard. Just when they had composed their faces for sympathy with the otters and waterbirds, someone stood up to cross-examine them angrily on their research on aquifers. Precisely which of "their" thirty-seven salmon creeks were they going to protect? Who was their riparian biologist, and what were his credentials? And his phone number would be nice.

Just as we were growing weary with this game, a young woman stood up at the back of the hall. She was tall, lithe, utterly beautiful, and looked at least part native, with long, dead-straight black hair, a weathered suede jacket that nonetheless draped gracefully on her frame, and a wide-brimmed, black felt hat with a band bejeweled in turquoise.

The loggers froze. The residents turned and craned their necks and, from the questioning murmur that arose, I guessed few knew who she was.

"Many people all over the world . . ." She paused, then repeated herself, her voice clear and strong. "Many people all over the world treasure this place and hold it sacred. Here and now I warn you. If you do what you are planning to do, you will stir up opposition that will cost you hundreds of thousands, if not millions of dollars. People will come here from all over and camp in your forests—thousands of them—until you leave. You will suffer. Your shareholders will suffer. Your company will not recover. So I tell you again. Leave now."

If you thought it impossible for predatory capitalists who clear-cut forests to turn pale, you'd be wrong.

People leaped to their feet and moved toward the three microphones set up for the audience's use, like excited shoals of fish. Shoulders slumped at the head of the room, the buzz became triumphant, and I had found something more interesting than city life.

I caught up with Emma as she left the hall, and she was just as exhilarated. The cause seemed just, the forest irreplaceable, and the land savable. Over the following year, parties and dances and sit-ins proliferated, the latter consisting mostly of whimsically dressed middle-aged islanders standing around on a dirt road in the middle of the day, fulminating. Fund-raising drives took place at every one of the many, many clubs and organizations on the island. I wrote columns about the fight for the national newspaper, the *Globe and Mail*, until my editor said he would neither publish nor pay for another.

Emma stickhandled the production of a calendar of nude island women to publicize the loggers' crimes against the forest and used her ex-husband's name to promote it. The island and the fight to save its forest popped up in newspapers all over the world. Protests and direct actions plagued the company until finally, a year to the day they arrived on the island, they gave up and sold the land to the government for considerably less than they had planned to make. The islanders had, despite the wool and patchouli and wood smoke, managed to raise more than $20 million—out of bloody thin air, as far as I could determine. And the loggers vanished, leaving, like chastened children, sprigs of planted trees in those sections they had managed to cut before hell rained down upon them.

Salt Spring and its four hundred sister islands big and small are run, for the most part, by a land-use outfit called the Islands Trust, the much-quoted mandate of which is "to preserve and protect." Founded in a summer graduate seminar at a local university in 1974 out of a fear that the pastoral islands scattered between Vancouver and Vancouver Island were about to be overrun by developers, the proposal for it was rushed through a session of the provincial legislature and its structure codified. The trust practices what is called fortress conservation, the typical form of conservation everywhere, which involves locking down as much land as possible and practicing "natural regulation," meaning no one touches it. Ever. Even "disturbing vegetation" is disallowed.

For most of recorded history, humans have practiced adaptive management of resources—when a problem crops up, we solve it. If we want a landscape, we create one; a working forest, ditto. Rangeland, farmland, townscapes—all can be managed for bounty and health of resources *and* people. Natural regulation cropped up in the 1960s in almost all land-use agencies in the world and swiftly became the preferred method by which all resources and land were to be managed. Over the past five decades, natural regulation has been adopted almost everywhere. Nature knows best. Man is a virus and a despoiler and must be controlled.

The Islands Trust went on to serve as a template for many similar organizations all over the world, including the California Coastal Commission and the Cape Cod Commission. The trust claims it is unique, but it is not, at least not anymore, and like its fellows, it differs from typical democratic government principally by subverting the normal processes of democracy in the name of good green land use.

There are thirteen larger islands in the trust area, to which the smaller ones are attached for administrative purposes, and those larger islands elect two trustees each. When I say *big*, it's relative. Most of the twelve other islands have a population between four hundred and one thousand, but they each have two trustees. An off-island trustee comes in for the monthly town meeting to vote, breaking any tie. All land-use decisions are first voted on on the island and then considered at a quarterly meeting of all twenty-six trustees, called the Trust Council, which moves with all the glacial formality of the League of Nations. There is, you have no doubt gathered, no proportional representation at the trust. Salt Spring has only 8 percent of the final vote on the way any of its land is used, and in rural areas, land use is just about everything.

The trust is a blue-chip organization; it is expensive, headquartered in British Columbia's capital city, with forty-five full-time employees and a steady flow of consultants. From its offices streams an unending flood of glossy propaganda about ecosystems saved and dangers advancing that require more land to be saved and more regulations placed on private land, and of course on all waters, whether runoff ditch or ocean.

Every few years, each island puts itself through a revision of its Official Community Plan. Carefully selected islanders serve on committees examining each "problem" within the trust region: affordable housing, tourism, economic development, water. Environmental movement goals are codified and tested during these thrash-fests of "participatory democracy"—goals like limiting house sizes to less than three thousand square feet, for instance. Or requiring a permit and the consultation of a registered environmental professional ($2,500 fee to be borne by the applicant) to plant a garden within 100 feet of any body of water, man-made or natural. Or requiring a 150-foot setback from the ocean for any house, and if a house already within that 150 feet burns, it cannot be rebuilt in the same place—well, tough luck for the stinking-rich oceanfront homeowner who just met Nemesis.

The outcome of a year or so of such meetings, displays, and "community consultation" is an astonishing maze of regulation, the result of which is stasis and worse. Despite living within easy reach of three gleaming modern cities filled with active wealthy-ish men and women, with the exception of Salt Spring, which is graying rapidly, every island's population is in steady decline, losing young families every month. On the smaller islands, as the young leave and the economy deflates faster, even the elderly, deprived of the services that the young provide and fund, leave too.

While the trust is supposed to deal only with land use, in typical bureaucratic mission creep it calls itself a local government. But in fact, all the other multiplying details of actual government are handled by a regional director, who spends his life dashing from committee to committee, all staffed by volunteers. Liability is limited, therefore, which may not mean much to the average citizen and in fact rule by volunteer committee sounds quite nice and inclusive and participatory, until those volunteers have mismanaged the drinking water so entirely that the lakes are poisoned with toxic algae, and no one is responsible—whereupon chaos reigns and the quarreling is epic. Salt Spring is typically described as an argument surrounded by water, as, I was to find, is every other community into which the movement has inserted its brand of land-use management.

But the trust and its islands remain an ideal, a banner brandished high by the environmental movement. Our islands are your future, in many ways, especially if you live in the country, but also pretty much everywhere, in the suburbs and in the cities where smart growth, an urban variation on the ironclad regime we endure, is rapidly being effected.

While I had not grown up on the Pacific coast, I did have roots in the region. Both sets of paternal great-grandparents had pioneered there in the 1870s, both of the men, curiously enough, named John Joseph. John Joseph Banfield was civic minded, and his family are still embedded in the life of Vancouver. The Nicksons, by contrast, never joined anything. John Joseph Nickson is described in the record as a "pioneer contractor," who laid out, then built waterworks and sewers—which flooded every year because he hadn't accounted for the torrential rains. But that didn't stop him. Harbors were dredged, neighborhoods clear-cut, sidewalks laid, tunnels dug, and terminuses, bridges, and breakwaters built in an explosion of enthusiastic late-Victorian growth.

The rest of his family, a wife and seven children, embraced a splendid isolation; according to my father, they were known for their beauty but mostly for their eccentricity. They became so peculiar that Vancouver society traded sightings of the last of John Joseph's children far into the 1980s. My great-aunt Rena, white braid hanging to the middle of her back, torn suede jacket, face resplendent, furiously strong, darting into a convenience store, would inspire days of conjecture. An Olympic rider, she had married a duke's youngest son and started pony clubs in Canada, but she lived in what can be charitably described as a hut, wherein everything was covered in generations of dog hair. Twenty paces from her front door stood her once well-appointed stables. In winter, she'd pack up her ancient beasts and ferry them to her farm on Vancouver Island, which was ever so slightly warmer—for the horses.

Rena's brother, Harold, lived up the coast on a thousand acres of waterfront, alone in the now dilapidated family house, a genuine hermit, his siblings' six bedrooms lined up on the water side of the second-floor hallway, kept as if they were coming back tomorrow, except for the grime on the chests of drawers. On his mile of beachfront, small cottages tucked into the trees were missing doors and windows; they had been guesthouses, married children's cottages, and playhouses, but all were long abandoned. He died as he lived, alone, pitched forward in the bed of his pickup truck, not found for three weeks, caught by death in his last ecstasy, I like to believe.

So while I knew next to nothing of forest or fish policy and even less about land use and conservation, I understood the ferocious euphoric pull of the Brobdingnagian rain coast. Greenpeace was founded here; in fact, this region produced the fiercest eco-warriors of the last generation: Sea Shepherd, Earth First and its demon spawn, the Earth Liberation Army, the Earth Liberation Front, and the Earth Liberation Institute were birthed here, their members captured by the scale of nature that so blandished then conquered my elders. And out of this region, forgotten by the rest of the world, came the ideas and the muscle that created an ideology and a strategic plan that to the average bear sounded so good, so positive and life affirming, that we all bought in. And then we watched as those ideas slowly brought the modern economy to its knees.

Like everyone, I was oblivious of all this until my land partner needed his money. An industrial filmmaker in San Francisco, Jim was pedaling his stationary bike one morning, shouting into the phone, and was felled by a stroke at thirty-nine. After two years of rehab, he was barely able to function, and he needed his money out of the land. We had agreed, in London a decade before, that he would retire to the little forest he'd help me buy, and we would each have a house. This was now out of the question, and after two years, his investments had dwindled to nothing, and he was facing desperate times. I had been paying him slowly, but the land's value had tripled since we'd bought it, so no payment from any income I could earn would ever pay him off in time. There was a solution, but it was so difficult to bring about that my mind shied away from it until it was clear that I had no choice.

A "density," in zoning speak, is the right to build a house. One density equals one house. Salt Spring, back in those heady times when the world was thought to be easily remade, had been granted 4,700 densities by its citizens' committees, divided up in various lot sizes. The average density would be one house per five acres. In order to create neighborhoods like mine—a typical rural, "medium density" neighborhood—a dozen smaller, 3-acre lots were created and surrounded by small holdings ranging from 10 to 160 acres. My 28 acres was a "remainder," which meant that it was left over from dividing off small lots, and was held as a unit to make up the 5-acre average. That meant it was treated as a 5-acre lot and I could build only one house. Which I had built.

However, I could buy a density or development right from another landowner on the island who, with his eighty acres, say, had the right to build four houses, and transfer it to my lot, thus giving me leave to build two houses, then subdivide. This much was legal. The trust wanted to move densities into the center of the island, where I lived, so that the outer forests and highlands would be held pristine. They approved of density transfers in principle.

In practice, however, they did not. Not one had ever gone through, of the hundreds that had been tried.

Although its pet bureaucrats had interjected transfer of development rights into land-use codes, the environmental movement disapproved of them, largely because they meant that the right to build on a plot of land where only a rich idiot having a nervous breakdown would try to build a house—water access only on the side of a cliff, say—would now suddenly be transferred to a central zone. A house that wouldn't have been built now suddenly could be. It seemed that instead of 4,700 houses, the most vocal islanders wanted only those that existed, 3,000.

This is called downzoning, and it is the movement's dominant imperative.

I needed to measure the extent of my problem, so I called up Emma, and we went off to a trust meeting. The room at the Lions club—a room with which I was to become depressingly familiar—was filled with people, all simmering with rage. But this was a different crowd. These people seemed older and more serious. And more conventionally dressed. A smattering of the patchouli and wood-smoke crowd was present but wasn't dominant. As the trustees took their places at the front of the room, the murmur rose sharply, then sank. Again there were no empty seats. Emma and I pulled chairs out of the storeroom and set them against the back wall.

The fug settled in, and the local trust committee settled down to muttering. Various announcements were made and agenda switches announced. The fellow I imagined to be the chair, a short, rabbinical type with a round face and dancing, amused eyes, then made a brief speech advising us to show the decorum for which the islands are famous. A few chuckles greeted his allusion: the islands are proud of their distinct lack of decorum. The meeting promptly devolved into a town hall segment, in which a great many people objected to a great many things, not one of which made sense to me. Floundering, I leaned toward Emma and asked, "Who's up there, anyway?"

"Small guy in the middle is the chair of the whole trust, David Essig. He's good, he's all right, he's one of us—he lives on Thetis. Doctorate from the London School of Economics—an economist." We nod at each other, impressed. "The two flanking him are the local trustees. The guy is Eric Booth, he used to be a *Realtor.*" The word was hissed rather than whispered. I note the long blond hair caught in a ponytail, intense blue eyes, workout clothes. He is staring crossly into his laptop screen. "The second, Kimberly, is all right, we think," Emma continues. "Union, gets it, on the whole, but not predictable."

"What about all the others?"

"Staff," she said.

"Why are they in such a foul mood?"

"Would you want that job?"

The dominant project under discussion was a proposed subdivision about twenty minutes from the village. The developer already had the right to build fifteen houses and fifteen guest cottages. He wanted to build twenty houses and no guest cottages and donate a large chunk of the acreage to the neighborhood as a park.

The neighbors were very much against the whole enchilada.

They did not like the traffic, they thought the park was too small, and by the way, they were used to having all of that land for their own. They did not see why houses needed to be built up there in the first place, and they cited water shortages, eagle trees, red-listed species—snakes and frogs mostly—the health of the residents and neighbors, noise issues, septic issues, possible rentals of these houses to party people, overuse of water by these people, the need for a vacation home rental moratorium, and while they were at it, the need for a moratorium on density transfers as well. I scrunched down in my chair.

A proposed density transfer to a property at the end of the island was now, in fact, under discussion. The arguments against this transfer were furious and went on for a very long time. Finally, a large red-haired woman stood up, and beside me Emma tensed. Sneaking a look at her, I saw that her face wore a mischievous smile.

In a large, stern voice, she barked a warning from West Coast Environmental Law, the dominant environmental law firm in the region. Apparently the Farmers' Institute owned a tiny sliver of waterfront land—less than five hundred square feet—near where the island's marina operator wanted to put in a chandlery and showers for boaters from Seattle and Vancouver. The farmers, apparently, had been seized with the imperative that no such thing should happen, and were willing to go to the mat to prevent it.

I knew Emma had recently been elected to the board of directors of the Farmers' Institute—"For the testosterone, darling," she had said.

"Is this you?" I whispered.

" 'Course not."

I didn't believe her.

Not one of the hundred people in attendance spoke in support of any of the three proposed developments. Yet each was legal and rigidly conformed to the land-use bylaw. I was to find out later that week, after an afternoon spent in the trust's office wrestling with five-pound binders of meeting minutes and letters from the public, that each of the three applicants had drastically altered his initial plan, going far beyond the requirements set out in the island's bylaws. Yet all three looked to be on the verge of failure.

I stumbled out of the hall with a blinding headache.

After deciding to fight, I took myself off to the local land surveyor—grizzled, despairing Brian Wolfe-Milner. After forty years working on Salt Spring, Brian knows more than anyone about the bones of the island, its zones and regions, and the regulation attached to each. I had gone to him as a reporter a few times, intimating that someday I might have to subdivide when Jim moved to the island. I stayed on to have several conversations, all ending with me nose-deep in despair while he watched with doleful amusement the process of discovering that the land into which you have sunk your savings is not really yours at all, all too familiar to him.

He alluded to the mountain of bureaucracy I'd need to scale, because the density transfer was only one aspect, the most difficult perhaps, but only one of a couple dozen others. I told him I had no choice. He contemplated me for a few moments, then recommended a colleague in the nearby town of Duncan, Brent Taylor, who might be willing to shepherd me through the process for $110 an hour. I was impressed both by the precision with which Brian knew the fee and by its size.

A few days later, Mr. Taylor heaved into my driveway in a dark green Chevy Suburban and squeezed himself into my little house. He was large and bearded and carried a stack of papers.

The beard covered the double chin of a man who had to work at not running to fat; the glasses hid eyes that I was to learn could beam an empathy that was almost physical.

What I wanted him to say was, "This is a war, and we are now calling this the war room."

Instead, he pulled out an application form, we filled it out (or rather, he did) in careful nicely rounded letters, and I gave him a check for $4,000. He smiled at me beatifically, with compassion, as if he were about to perform an operation and that operation would be painful and last a really long time and he didn't have much in the way of painkiller to offer. I knew I had to ask, but was putting it off, dreading the answer. He was pinning the check to the application form, slipping it into a manila envelope, not in haste exactly, but he was trying to get out of the house.

"How long?" I asked. The files stopped disappearing into his briefcase.

"I'm sorry?" But he knew what I meant.

"How long till it's done?"

He shrugged. "Hard to say."

"Try," I said.

"Nine months, absolute minimum," he said finally, unwilling to the last breath. He thought for a moment. "But that's if everything, and I mean every single thing, goes your way. Which it never does. And you'll have to change the political system."

Actually he didn't say that last sentence. Nor did he bother mentioning that the trust had not yet approved any density transfer. But he did say it was under no requirement to even accept my application *into* the rezoning process. That one step itself might take months, and there was nothing I could do but wait. Oh, and his other client? Almost three years into the same process, with no end in sight.

• • •

I drove down the hill to see that client. Janet Unger owns a hard-scrabble nursery and lavender farm at the end of the Fulford Valley, tucked into a hollow that looks as if Frederick Law Olmsted gazed down from heaven and guided God's hand just for that moment. Unger is an Englishwoman of a recognizable type; they pop up in every farthest-flung pretty place, hacking sensible civilization out of solid rock. When I drive up, an elderly man with a pickup is helping her pull her massive wooden sign out of the ditch and reposition it. The nursery is filled with plants in various states of winter decay. A fifty-foot-long pergola has collapsed and a tin roof has buckled under the weight of the recent snow. She and her husband own a house on fifteen acres on an outlying spit, and it is that property that she is trying to rezone and that had elicited the most fury at the meeting the other night.

We take grocery store plastic bags around to the back and pat them down on plastic chair seats to protect ourselves. Everything is wet. In front of us rises a sheer face of green, the tree-covered mountain. Between us and the mountain lies a meadow, green again after the last spectacular blizzard. It is beautiful, but neither of us is in the mood for contemplation. And Janet Unger has no more time to waste.

"It was so personal," she said. "So vindictive. They kept saying it wasn't personal, but it was. Something that should have taken nine months has taken more than three years."

I ask who "they" were. "My neighbors," she answers. "We lived on that property on Isabella Point for almost fifteen years. They were our friends; we played bridge with them every week. We had a nice house on the ocean, about a twenty-minute drive from here, with fifteen acres of forest and rock face, uninhabitable land mostly, surrounded by thousands of acres of park and ecological reserve. One of our neighbors, our best friend, in fact, when we lost our retirement money in the stock market, suggested we do a density transfer and put in a four-house subdivision. He and his wife had done a similar thing a few years back, and he was a developer himself, so he knew what I should do."

I can guess what happened next, but I ask anyway.

"An orchestrated campaign began, attacking me and my husband as despoilers of the environment. Letters to the paper were written, and person after person stood up at meetings to tell the community how terrible we were to try to do such a thing. Despite the thousands of acres of wilderness around us, they were insisting on a buffer zone for that wilderness. We had applied to build four extra houses, so when that application was turned down, we resubmitted it for just two additional houses on the fifteen acres. That took another twenty months. You can't believe how brutal they were, the lies they told about us. We couldn't look our neighbors in the face as we drove by. Our best friends deserted us. It broke up our thirty-two-year marriage; we couldn't speak to each other by the end of it. Last year, my health collapsed; I have cancer. If I live, I won't live in that house in that neighborhood or on this island. I'm selling everything, including the nursery."

I am cold, but it isn't the weather. "Who started it?" I ask. There are only a few candidates, but I need to know which, in order to try to neutralize her. And it will be a her.

She shrugs and mentions a charismatic local environmentalist as the source of the attack. "Though maybe now," says Janet, dry as paper, "she has changed, since she is developing her own family's property, subdividing and selling two oceanfront lots for a million dollars each. All her permits are going through lightning fast, I can tell you."

"Get prepared," she says over her shoulder, on the way back to fixing her sign, "for people to hate you without reason."

• • •

Today, Janet Unger lives in a trailer park. It is a groomed and manicured trailer park, where most of the pretty little houses, modeled on the double-wide, are set on concrete pads, not wheels. But it is a long, long way from the five-bedroom shingled house on a cliff overlooking the ocean. Her rezoning and subdivision took seventy-two months, and by the end of it, her marriage was dead, her husband was dying, and all her money had disappeared down a bureaucratic wormhole.

Her nursery stands abandoned. By the time she was able to put her property up for sale, the market for country real estate had declined and then, with the 2008 crash, collapsed. She was lucky to afford a little house in a trailer park.

The environmentalist who led the charge against her went on to better things. Her family's subdivision had sailed through, and she built herself an ultragreen house with ever so virtuous salvaged materials and settled down to work for outfits like the Western Climate Initiative as an independent woman.

The Western Climate Initiative is a collaboration among seven U.S. states and two Canadian provinces to design a regulatory regime to reduce greenhouse gases. This is a typical career progression in the movement. A "kill" of someone deemed to be a developer—even if she is only dividing fifteen acres in thirds because she's lost her retirement money—is a required initiation; it proves you have the stomach for what needs eventually to be done.

Transnational organizations are typical too. They sound high-minded and entirely good. Their final goal is the power to regulate your every exhalation, for what else is it but CO_2? It is axiomatic that no voter will know what's happening until it's too late. Democracy is always and everywhere overridden when it comes to green goals.

Little doubt, too, that Janet's neighbors have not counted the cost of their ever so righteous crusade on behalf of the environment (and the exclusivity of their neighborhood). No doubt they believe ruining her family was a terrible necessity that would serve as a stark warning to any purveyor of ticky-tacky condo developments for the great unwashed.

The bureaucrats who oversaw the seventy-two months of Janet Unger's misery were not held accountable for the bust-up of her marriage, her husband's fatal heart attack, or her cancer. Her bankruptcy at their hands meant nothing to them at all.

CHAPTER THREE

The Big Big Picture: Just How Much Land Has Been Saved, Anyway?

In 1788, my great-great-great-great-great-grandfather, Oliver Phelps, pulled off what still stands as the largest private real estate deal in American history. From the bankrupt Massachusetts legislature, still reeling from the cost of the Revolutionary War, he bought 6 million acres of what was to become western New York State. The million dollars ($20 billion today) he promised was essentially the first debt-for-nature credit default swap, a contemporary mechanism that has laid waste to the developing world. Phelps and his partner, Nathanial Gorham, bought Massachusetts's debt, which was heavily discounted at the end of the war. In its complexity, aim, process, and results, his arrangement bears a striking similarity to the sublimely clever, post-postmodern land acquisitions of the environmental movement.

Typically the big international land trusts buy Third World debt (or have it donated) from American banks who decide to go for a tax deduction rather than payment, of which they have long despaired. The ENGOs then go to the developing country whose debt they now own and offer to retire it, on condition that certain biologically valuable tracts of land in that country are "conserved." In order for those lands to become "pristine wilderness," they must be cleared of people, of course. New villages are promised by the ENGO and the host country, with proper hygiene and green jobs, windmills, and solar panels, but in a striking similarity to the promises made to North American Indians, that idea dies on the vine.

An energetic smooth talker, Phelps had more confidence than ten men, which is probably why he died in debtors' prison. After his thrilling maneuver in Boston, he promptly moved to the foot of Lake Canandaigua and built himself an expansive shingled saltbox (today the Bed and Breakfast at Oliver Phelps), established a land office, and began surveying. Within a couple of years, he invented a procedure for dividing land into ranges, counties, townships, and sections, a system that became the basis of county and town platting for new territories throughout the United States as frontiers opened up in the decades following the Revolution. His platting underpinned county planning until the 1970s, when green land-use regulation began. The Phelps-Gorham Purchase kicked off a shift in the way people lived on and off the land, as system changing as the transformation we are undergoing today.

As a member of a Puritan family who had prospered in New England since the 1630s, Oliver Phelps understandably fell into being deputy commissary of the Continental Army. It was also understandable that when George Washington traveled into what would become upper New York State after the war, Phelps was there to "help," especially in promoting the idea of saving the wretched land-poor peasantry of Europe; Phelpses, while they liked making money, really really liked good works. Washington fell in love with upstate New York; here was "the seat of Empire," he declared. Let this lush and fertile terrain, pockmarked by hundreds of rivers and lakes, be the foundation, the future, of the new republic. Washington, DeWitt Clinton, Nathanial Gorham, Robert Morris, and all the other great projectors of the American republic genuinely believed that the postrevolutionary reconfiguration of America's land was their mission from God, in order to rescue their brethren in Europe,[1] locked in quasi-bondage or millennial poverty.

Private ownership was key. Plymouth Colony had begun as a communal enterprise, but it was not until each family tilled and profited from its own plots that the colony began to thrive, or in William Bradford's words, brought "very good success." A man who owned enough land to feed and clothe his wife and children without becoming a mendicant in the marketplace was capable of standing up to tyranny and in fact had just done that very thing. To Washington and his associates, the independent yeoman farmer was the unbreakable unit of American freedom. Charitable as those drenched and marinated in Christianity can be, it was the rescue of their fellows—the extension of the American dream—that gave the Revolutionary generation the holy purpose that justified what they would do next.

With his "seat of Empire" statement, Washington decreed that the 25-million-acre ancestral territory of twenty thousand Mohawk, Oneida, Tuscarora, Seneca, Cayuga, and Onondaga Indians was now open for business. No doubt they all believed that a density of one Iroquois per 1,250 acres was an inefficient use of land, but Washington's subsequent actions set off a simmering hatred that lasts to this day.

New York State and Massachusetts immediately claimed Iroquoia as their territory. New York got the land and Massachusetts, the right to preempt Indian title. That way, New York State would be able to tax the land in perpetuity, and Massachusetts could sell it to Phelps and Gorham for a considerable amount of money. In order to ensure ownership, Phelps and Gorham had to buy the Indians' title. In effect, they bought the land twice. As soon as the legislature finalized the deal, Phelps heaved himself up on his horse and rode out to meet the Indians who had lived on their land for a millennium or so and were about to be deprived of their ancestral homes. Phelps somehow managed not only to talk the Iroquois out of scalping him—they wanted to—but also into selling him their title. The money helped. It wasn't a bad deal, but what happened next wasn't pretty. A brutal winter set in and the Iroquois crowded around the border forts on their way to Canada, which had set aside land for British Loyalists, red and white. But the fort was choked with emigrants, and many froze, starved, and died. Any payment for the Iroquois vanished when Phelps went bankrupt. Massachusetts debt was suddenly worth real money in the boom times that followed the Revolution, and he lost the land and pitched up in debtors' prison, and not for the last time either. Subsequent owners did not honor his contract with the Iroquois, and in a repudiation of the Revolutionary generation's treatment of the Indian, the Phelps generations that followed became furious and vocal advocates for the Mohawk,* a rejection that surely waits for today's conservationists.

In an eerie parallel, the Boomer generation, with its self-righteous freak flag flying and its wholesale condemnation of our white ancestors' treatment of the Indians of North America, is effecting clearances far greater in scale than that first, state-sponsored clearance of the American republic.

Mark Dowie, a former publisher and editor of *Mother Jones* and winner of fourteen National Magazine Awards, found, after two years of travel into far-flung jungle, upland forest, and bottomland, that more than 14 million indigenous people had been cleared from their ancestral lands by conservationists. His journey is chronicled in *Conservation Refugees: The Hundred-Year Conflict Between Global Conservation and Native Peoples*, published

* One of Phelps's nephews married into Joseph Brant's family on the Five Nations reserve in Ontario, and became such an avid advocate for the Mohawk, he had to be smuggled out of the country to Chicago, because the British were about to hang him. Equally, Phelps's grand-nephew, O. S. Phelps, wrote florid Victorian tracts about the white man's mistreatment of the Indian.

by MIT Press in 2009. Both before and after Dowie's travel and his deeply researched peer-reviewed work, scholars in the United Kingdom and the United States data mined for the number of persons displaced by conservation, placing those numbers upward of 20 million; some analysts claim there may be many millions more.[2] The World Wildlife Fund, the Nature Conservancy, and the UN-affiliated Conservation International are the principal players, and while they do not directly clear people from their homelands, they step aside when national governments do so, at their request. The conservators maintain plausible deniability and assert that they are creating "sustainable" economies, proffering examples that serve as today's version of the Potemkin village.

> *"Men in uniform just appeared one day, out of nowhere, showing their guns," Kohn Noi [matriarch of sixty-five families of the Karen, one of six tribes found in the northern Thai forests] recalls, "and telling us that we were now living in a national park. That was the first we knew of it. Our own guns were confiscated . . . no more hunting, no more trapping, no more snaring, and no more 'slash and burn.' That's what they call our agriculture. We call it crop rotation and we've been doing it in this valley for over two hundred years. Soon we will be forced to sell rice to pay for greens and legumes we are no longer allowed to grow here. Hunting we can live without, as we raise chickens, pigs, and buffalo. But rotational farming is our way of life."*[3]

In this 2004 case, the Global Environmental Facility* lavished money on the Thai government, encouraging it to set aside three forest reserves. The country created 114 new parks, adding

* The Global Environmental Facility (GEF) serves as financial mechanism for the following conventions: Convention on Biological Diversity (CBD), United Nations Framework Convention on Climate Change (UNFCCC), Stockholm Convention on Persistent Organic Pollutants (POPs), and UN Convention to Combat Desertification (UNCCD).

up to more than 9,500 square miles, in a matter of months. The ongoing parallels with the early American treatment of the Indian are eerie. Just as the white man installed Indian chiefs with whom they could barter, so the Thai government and conservationists elected chiefs from clans whose governance had never called for such rulers. Nor is this a dirty little secret; critics within the movement find such behavior repellent.

In 2004, Mac Chapin wrote in *World Watch* magazine, a movement publication:

> *Around this time, a CI [Conservation International] biologist who works with the Kayapó in the Lower Xingu region of Brazil told me: "Quite frankly, I don't care what the Indians want. We have to work to conserve the biodiversity." This last comment may sound crass, but I believe that it accurately represents the prevalent way of thinking within the large conservation organizations. Although they won't say it openly, the attitude of many conservationists is that they have the money and they are going to call the shots. They have cordoned off certain areas for conservation, and in their own minds they have a clear idea of what should be done. "They see themselves as scientists doing God's work," says one critic, pointing out the conservationists' sense of "a divine mission to save the Earth." Armed with science, they define the terms of engagement. Then they invite the indigenous residents to participate in the agenda that they have laid out. If the indigenous peoples don't like the agenda, they will simply be ignored.*

The result? Millions now live in shantytowns ringing their former homelands in Africa, South America, and Southeast Asia, angry, hungry, and so disoriented that there are reports of some walking blindly into traffic.[4] The payoff: in the last ten years, the number of international parks and protected refuges in the developing world has doubled, to more than 108,000. More than 10 percent of the developing world's landmass has been placed under strict conservation—11.75 million square miles, *more than the entire continent of Africa*.[5]

Everyone conserves in the developing world, setting aside trillions of dollars in natural resources. In 2010, Goldman Sachs announced that 735,000 acres had been placed under conservation in Tierra del Fuego, the economy of which is based on its land: oil, gas, sheep farming, and ecotourism. You can be sure that these 735,000 acres were some of the most valuable acres in Tierra del Fuego. *All* set-aside lands hold valuable natural resources that could have been used to create a landed peasantry, thereby moving natives out of a millennium of structural poverty, or to build hospitals, schools, and a cheap energy system for all the citizens of that developing nation.

Little wonder that country people see themselves as the new persecuted class. We haven't even begun to count the tens of millions of crushed and broken rural Americans cleared from the country, the resources shut away by the beneficent rich, the opportunities lost, the lives choked, the children who could not be sent to college, or the unfunded retirements. The subversion of one of the founding principles of the American republic—private property—demonstrates just how productive that principle was, creating prosperity for hundreds of millions. Today's proliferating restrictions are creating disaster for tens of millions. Collapse is easy to engineer once you effectively own all property. Environmental overregulation is what, in part, motivates the Tea Party.

In the United States alone, hundreds of millions of acres—more than 30 percent of the nation's land area—have been set aside in formally restricted zones, whether wilderness, forest reserve, or privately conserved land. According to Conservation Almanac, a website of the Trust for Public Land, as of 2005, 20 percent of the United States, or 473,653,970 million acres, had been placed under no-use or limited-use restrictions.[6] As of 2010, the count is up to 700 million acres, one-third of the U.S. land base of 2.3 billion acres.

In Canada conserved or severely restricted lands are much more than 30 percent. Most of Canada's land is owned by its government; in British Columbia, only 6 percent of the land is privately held, and only 1.9 percent of that is available for development.[7] In fact, little more than 3 percent of the landmass of Canada has been developed, yet in late 2010, the country placed 150 million acres of boreal forest, an area twice the size of Germany, under strict conservation, except for those multinational forestry companies who have made deals with the American foundations and multinational ENGOs that now oversee those lands. Ownership, for all intents and purposes, has been transferred out of Canadian hands. In February 2012, the premier of Quebec, Jean Charest, announced what he called perhaps the largest conservation project on the planet, an area the size of France, roughly 160 million acres.[8]

It is useful, therefore, to demonstrate what happens in a region when conservation becomes the dominant value, so for the sake of argument, we should look at the state where the American "empire" began, keeping in mind that the third most populous state in the union, New York, in the popular imagination is generally thought of as heavily industrialized, populated, and polluted, in the greatest danger therefore of catastrophic biodiversity collapse.

This is so wrong as to be ridiculous. In fact, more than 40 percent of New York—almost 14 million acres*—is in the process of being rewilded, turned back, in all essentials, to Iroquoia.

* Acreage counts are slippery, especially when you are counting millions of acres. As land is surveyed and property rights asserted, acreage counts will shift.

Ten percent (more than a million acres) of New York State is classed as wilderness, and another 20 percent (more than 2 million acres) is classified as forest preserve, constitutionally protected as "forever wild." Another 2.8 million acres is formally protected and managed by the government in various categories, adding up to 5,804,000 acres in government-managed landholdings. Acreages are not published for the 1,349 miles of wild and scenic rivers, coastal erosion hazard areas, coastal fish and wildlife habitat, and the buffers that these treasures require. Nor are the 1,251,632 acres of wetlands counted in protected areas; nonetheless, these areas are protected, and all of them have 100- to 300-foot buffers around them. Equally of the 70,000 miles of creeks and streams in New York State, all are in the process of having buffers instituted. Wetlands in Adirondack Park add another million acres of protected wetland; the state Department of Environmental Conservation has not as of this date finished the count of protected wetland within the park. Depending on the size of the wetland, buffers can take many more acres than the wetland itself. But a rough calculation that places buffering of creeks, streams, and wetlands using a conservative 200-foot buffer (the state DEC states it wants 300-foot buffers) increases land set-asides to another 4 million acres.

The template for "living embedded in nature," the movement's stated, overarching goal, is demonstrated by the fate of those living in Adirondack Park under "ecosystem management," which is the standard toward which all rural lands are heading. That tight management, allowing one house per 8.6 acres or one house in 45 acres, brings the count of acres under tacit preservation to 14 million acres.

Founded twenty years after Yellowstone, in 1892, Adirondack Park was originally 2.6 million acres. Today its acreage stands at

6.1 million acres, the same area, spookily enough, as the Phelps-Gorham Purchase. Almost 60 percent of Adirondack Park is privately owned; in fact, much of the park has always been private land. Today the park holds 103 small towns and hamlets; the 130,000 residents all live on privately owned land. As the steel net around those people has tightened, their lives have become a reversal of the accepted progression of modern life: poorer, with sharply limited opportunity and access, and older, yet with shrinking government services.

Many of the mechanisms of conservation—such as greenlining, the targeting of private land for conservation—were first invented for use in the Adirondacks. Unlike Yellowstone, it had been settled by early pioneers for more than 150 years when created. As a result, the park is the model that we are all, in one way or another, following in the creation of protected spaces that were once or are still lived upon or used by humans in some way. Management of the one hundred national monument sites in the United States, for instance, mimics almost exactly the management of Adirondack Park. In an expansion process typical with all conservation, the establishment of a staggering sixty additional monument sites awaits President Obama's signature. There is no procedural way to fight or stop a monument listing; it is entirely an executive decision, and once done, anyone with interests in that area has essentially been moved to the scrambling and desperate working class.

In the fall of 2009, the Adirondack Park Regional Assessment Project released a report describing the conservators' "success." In the past thirty years, school registration has dropped 30 percent, meaning that the equivalent of one school has closed every nineteen months. Those under thirty-five leave the park as fast as their legs can carry them, but retirees replace them, and the aging trend in the park is three times the national average.

Incomes are lower within the park, lower even than those towns only partially located inside the park. Mines and mills—as forests have been gradually moved into "forever wild" status—are shuttered. No private investor dares put money into the park; as a result, as late as 2009, only three of the 103 towns had cell phone coverage, and there was virtually no broadband. Jobs in the private sector have vanished, though until the 2008–2009 crash government jobs had increased, subsidized, of course, by the state, since local government is broke. Between 1986 and 2006, property tax revenues dropped 3 percent in the park—this during the biggest real estate and accompanying property tax boom in history. Municipal expenses—largely driven by welfare and the costs of complying with onerous environmental regulations—are twice that of towns outside the park. Typically, conservationists promise that with forests and lakes moved into wilderness or "forever wild" status, tourism will replace lost jobs. But that *never ever* happens.

In 2006, Carol LaGrasse, a retired civil engineer from New York City, and Susan Allen, editor and publisher of a local Adirondack newspaper, took a walk on the highway out of the town of Wells (2000 population, 737). The couple were headed for Whitehouse, a famous ghost town in the park, first settled, as was Wells, in the last years of the eighteenth century. Whitehouse, with its two antique suspension bridges and ancient chimneys, has been a well-loved recreation area for Wells residents, who consider the settler artifacts in Whitehouse part of their cultural heritage. LaGrasse and Allen found that the main road into Whitehouse was blocked by boulders and was scheduled to be further broken up into a footpath. The culverts were to be removed by the Department of Environmental Conservation. Old roads that led to campsites long used by the townspeople of Wells were now completely blocked. In the ghost town proper, the fieldstone chimneys were not long for the world either. "The chimneys are non-conforming under the APSLMP [Adirondack Park State Land Master Plan] Wilderness guidelines," trumpeted the DEC. Of course they are nonconforming. They were built in the late 1700s. Until it was stopped by the town supervisor (given the vocal opposition of all the residents of Wells), the Department of Environmental Conservation planned to tear down the suspension bridges and chimneys and break up the stable sod near rivers and ponds, established by settlers and now used by those settlers' descendants as campsites. Old grave sites would be inaccessible to those descendants who cannot hike in.

LaGrasse:

> *Eradicating highways closes down access to the cherished places, so that descendants and local people who would like to visit, pay their respects, and maintain the cemeteries find it impossible. More harshly, the normal effort of people to show reverence toward those who came before them is squelched. Deliberately closing down access to local cemeteries . . . becomes a demoralizing force in the human community. It is a force of tyranny and debasement.*

The Indians of North America complain of the white man building on their old sacred grounds. Nary a voice is raised against the legitimacy of these claims. When it comes to their own heritage, however, today's conservationists are more than willing to gut their past.

Thinking on property rights has become infinitely complex over the past 250 years, but it is arguable that no conclusion has advanced further than that of Aristotle, as long as you toss out his definition of anyone not Greek as a slave who did not deserve property: "What is common to the greatest number gets the least amount of care. Men pay most attention to what is their own; they care less for what is common; or at any rate they care for it only to the extent to which each is individually concerned." The shift to property ownership that became institutionalized in the years after the American Revolution changed everything.

Most critics of our metastatic planning and regulation like to trace its beginning to the vastly successful mechanization that won World War II. After that victory, public intellectuals thought that planning could remake the world in fairer, more productive ways. Propertied individuals could not be controlled so as to make them part of a truly effective plan. This was a perfect inversion of what Washington and his associates thought—that propertied individuals were free individuals. To postwar planners, profit as motive would have to be replaced by a higher set of values, which boiled down to service to the community at large. Certainly during our seemingly endless round of meetings on Salt Spring, the common good is held up as the virtue toward which we all must aspire. Luckily for us, there are dozens, if not hundreds, of Salt Spring citizens who are willing to show our unreconstructed selves just how to bend our lives to serving the community. All of it requires more planning, more studies, more regulation, and more confiscation.

In the wider culture, however, by the 1980s, central planning had failed so many times and in so many ways that it had been largely abandoned in embarrassment. In the developing and Iron Curtain countries, where there were no constitutional democracies and no private property rights, the planning ethic could be fully realized; the results had been shown to be catastrophic. However, lots of people felt that the higher value of community service (or Marxism) failed because it hadn't been tried using the bounty of the West. It has been said many times but bears saying again, when the Iron Curtain fell, fellow travelers migrated to the environmental movement. And when they arrived to transform the rural world—a world few of us visit except on vacation, when no one is paying attention—they brought their planning with them. And because no one has been paying attention, the planning and control of rural areas has been adopted in countries big and small, developed and not.[9] The shift from Phelps's ranges, townships, and counties to inch-by-inch biocentrist command and control is one that would pitch George Washington and his associates into a state of awe at its reach. They would barely be able to grasp the fact that these new values are supposedly international and are definitely coordinated and implemented in a systematic way in many regions at the same time. And they would certainly rebel at the control these new values demand. As Tom Bethell makes plain in his extraordinary examination of the linkage between private property and prosperity,[10] the constitutional governments that emerged after the American Revolution recognized that they did not have the right to interfere with the law-abiding activities of their citizens. And when that happened, says Bethell, the result was a tremendous increase in material prosperity in those countries that recognized property rights.

By the 1970s, this principle was looked upon as an historical truism that no longer held value for anyone, neither the so-called developed West nor developing countries. The mathematics and science of planning would prove to be so much more productive than ideology or principle; that was obvious, said economists of the time. But would it? There was no case any planner could point to that proved that planning worked, except for China. The slow development of India was often cited in contrast to rapidly growing, centrally planned China. It was thought that India's growth was slow because planners were not ruthless enough. Nehru himself was a "thought-leader" in this parade of foolishness. In fact,

Nehru asserted that underdeveloped countries *proved* that private property had been superseded by the planning state. Nehru said property was immoral, far more so than drink, because private property elevated the individual over society, and that was clearly wrong.

The founding principles of the United States, at least when it came to private property, worked only because the states were uniquely circumstanced. Things were different now. Bureaucrats could create prosperity with government money and enforce equity through planning, accompanied by surveillance and coercion. At the time, aid money to developing countries was thought to be the magic bean that would produce the beanstalk *and* gold coins. Aid money was accompanied by planning advisers who thought it was not strictly necessary to have the institutions of capitalism, principally private property and profit, constitutional democracy, and the rule of law, in order to create growth.

These ideas dominated both academy and governments. Joan Robinson of Cambridge said in the 1960s that "great inequalities of wealth" were necessary during the industrial revolution, but now, in the hypermodern superstate, property rights were "otiose." They served no purpose and should be retired. The State must rule with an iron fist because that was the only way to create wealth and equity.

There is no starker way to describe what is taking place right now in the country than as the full flourishing of the bureaucratic state. Private property rights have been largely removed, the culture is dying, but the state, consisting of federal, state, and local ministries and departments, has bulked out so that a giant superstructure of bureaucrats with rulebooks piled high around their desks flourishes, grows, and feeds on ever-diminishing wealth.

Tom Bethell makes clear that the planning state's first order of business as it came to power in the 1960s was to export its ideas to the developing world. Fifty years later, this culturewide dream lies in a heap of wreckage.

Despite the cataclysmic failure of the planning state in the developing world, property rights have still not recovered, not even in the country where they were shown to be *the* engine of abundance. Today, insofar as anyone has property rights, they are viewed as a bundle of sticks; for instance, the right to the water on that land is a stick, the mineral rights on that land is another, as are the rights to build one or a hundred houses on that land or to cut trees on it. Each stick can be removed, one or two at a time, by government claiming the land would be more profitably used in another way or is "at risk," that the public good would be served by removing density or water. But despite all the promotional materials to show how good and noble it is to "save" land, to set it aside or sell it to the government to be permanently protected, there is no economic model in history that shows that communal holding of land creates prosperity. Nor is the effect neutral. Communal property ownership always creates economic malaise or worse. Nonetheless, we are still working toward that model, under which the state owns everything and permits everything.

There are hundreds of ways to remove property rights without outright purchase of the land, and New York State pioneered the use of nearly all of them. Thirteen hundred miles of "wild, scenic and recreational rivers" are protected in New York State; try to locate a new dock, house, development, or other private enterprise near one of *those.* A partnership among the Natural Heritage Program, the Nature Conservancy, and the Department of Environmental Conservation to locate and map rare species and significant habitats is shaping up to be another way to shift private property to no-use status. Another partnership, with the National Audubon Society, will protect "important bird areas and bird conservation areas," many of which will be on private land.

The Open Space Plan has also been expanded to protect estuarine areas, dunes, bluffs, shoals, barrier islands, contiguous and associated saline and fresh waters, significant coastal fish and wildlife habitats, scenic areas, and benthic habitats. Try to find a hunk of desirable rural land that does *not* include water or scenic elements.

To make sure no piece of land goes unsaved, the ninety-three private land trusts in New York State all search ceaselessly for lands to acquire. In addition, the Hudson River Valley Special Resource Study Act, now in front of Congress, is the first sally in an attempt to lock down activity in the twelve counties bordering the Hudson River. Conservation history teaches us that the 3 million people and the 250,000 acres in those counties are probably headed for national monument status. The Hudson River, the commercial spine of the state of New York, the river that arguably began the creation of the world's economic powerhouse and most egalitarian city, is on its way to becoming a theme park, its people and rural communities sucked into a deflationary spiral.

As resource use is curtailed, jobs are extinguished, service jobs fall away, families break up, tax money vanishes, and social services fade just like the Cheshire Cat's smile. In rural counties all across the United States, sections of every forested county were once deeded as school trusts. The proceeds from the sale of the wood from those forests would be spent by the public schools in that county. As forests are reassigned to no-use status, rural schools lose that income and are gutted.[11]

Conservers acknowledge that removing land from the tax base means government services decline because property-tax collection crashes. Many states promise something called payment in lieu of taxes, abbreviated PILT or PILOT. But PILT never replaces all the tax revenue, and it usually peters out within three years, after which tourism and that mythic beast, green jobs, are supposed to take up the slack. In a final eerie parallel, just as the resettlement money that was promised the Indians of North America rarely showed up, Mark Dowie reports that resettlement money for tribes cleared from their ancestral land by the Nature Conservancy, the World Wildlife Fund, or Conservation International rarely arrives. Conservers in the developing world and in North America point to joint nature/resource projects. Regrettably, most of these fall into the "too little, too late" category, a cruel joke to country people and Potemkin villages for the world's press, corporate and government donors, and anyone else who asks an inconvenient question.

The human culture may die, but at least the forests are protected and healthy. Right? Right?

We Have Just Canceled Your Vacation

Back on Salt Spring, I attend a meeting on vacation home rentals. Instead of shades of subtle green slumped in repressive vigilance, a different group of people are here to plead their case. Their faces are muscular, and old and young simmer with energy. In the Salt Spring summers, many of them pack up, move to trailers, and rent their houses to tourists. Salt Spring is expensive, property taxes are high, and this maneuver is the only way many can afford to stay on the island. The Islands Trust has been instructed by the conservation community to shut down this activity. In other places, it intones, vacation home rentals have led to entire neighborhoods being gutted by transient party people. And although on Salt Spring such rentals have mostly attracted families with young children or couples looking for a country retreat for a week or two, the green lobby claims that more visitors deplete the water supply, create more sewage, trample the parks, clog up the village, shop too much, and increase vulgar, environmentally unsound commerce, and most important, that such profits will lead to everyone renting their house in the summer to make a quick buck. Vacation rentals must be criminalized immediately.

The two local trustees—working class and therefore sympathetic to those for whom an extra $5,000 a year is critical—have resisted the pressure to shut down the practice. While not specifically permitted in the Official Community Plan, summer rentals are not illegal. But after five years of lobbying by the green community and two more years of hands-on wrangling, eating up thousands of hours of staff time, a new bylaw controlling vacation rentals has been written and given to the community for discussion. Under this new bylaw, on an island with ten thousand residents and three thousand houses, precisely 200 permits for vacation rentals would be issued each year. This is not onerous, because only 179 vacation rentals had operated last summer. However, new restrictions and requirements would be extensive: you couldn't rent your house for more than thirty continuous days, for example, nor could the rental be in a watershed. You'd have to be a permanent resident of the island who "continuously" occupies the house. You would have to prove water potability, have a functioning septic system, provide extensive parking, and take responsibility for any excessive noise from your renters.

Several hundred people are present, and at the front of the room sit the two local trustees, Kimberly Lineger and Eric Booth. David Essig has come in from his much smaller island two ferry rides away. Each of the thirteen largest islands has two local trustees, and David travels around to each island's monthly meeting to serve as the third vote to break a possible tie. This is a deliberate tactic to control local trustees who might be inclined to rule in favor of their neighbors and friends rather than Mother Nature.

Three staff planners sit at the table. They keep their heads down, and when answering questions, speak in voices so low that people strain to hear.

I occupy myself by assigning fairy-tale identities to the public servants. David Essig is Toad of Toad Hall, small, erudite, and despite the twinkle, stern. Kimberly is a Little Rascal. She is forty-two, with a mobile face that grins with delight whenever her photographic mastery of the island's fiendishly complex bylaws is called upon. Alfalfa's cowlick springs out of her crown; Kimberly is an irrepressible fount of regulation.

A young woman stands up. "Will the bylaw be approved in time for summer?"

Kimberly shrugs. "These things take time, and you know how the trust works." There is a SLOW sign propped in front of her place at the table.

"We need to know whether we can plan to rent our house this summer and for how long, thirty days, sixty days, which?"

"I don't know," says Kimberly. "All will be revealed."

"We need to have time to prepare with all these new rules."

"All in good time."

"And if, which is a big if, I am able to pass all the inspections and I get a two-year permit, can I count on having my license renewed for the next two?" The supplicant points out that the income is critical and that, while she and her husband work, they have three kids, and prices are high on the island.

She does not mention the costs of complying with the trust's proposed bylaw on vacation rentals. She doesn't dare; she is on the verge of losing the right altogether. She merely refers to the costs in passing. Nonetheless, as a disincentive, the compliance costs—of testing water and septic fields, of building a parking lot, and of substantial additional insurance—may well exceed what the family can make in one summer, and everyone in the room is aware of that.

"We have to be able to make financial plans. The trust needs to recognize this," she says in conclusion and sits down, her face flushed. Personal humiliation seems to be a necessary part of the process, and this woman with three small children—busy, hard-working, useful—is angry and embarrassed that she has to beg.

Kimberly prevaricates, and Eric Booth, her fellow local trustee—he plays Scrooge in the annual production of *A Christmas Carol*; unfairly, the association sticks—attempts to soothe her.

The young woman stands up again. "One of our neighbors complains because our children make too much noise playing. If that neighbor fills the trust office with letters of complaint, will the trust be reasonable and allow us to continue our business?"

"All complaints are investigated, and few are found to be reasonable. And the trust is always reasonable," says Eric.

The audience rustles and mutters, and there is one loud snort.

Kimberly flares up at the sound. "If anyone, anyone at all, rents their house before the bylaw receives final reading, the RCMP will be called."

Kimberly's day job is with the Royal Canadian Mounted Police, so a frisson of terror passes through the audience. How is it possible that something families have been doing for decades is suddenly illegal?

There is a way of killing things in the forest. The hunter places his boot on the heart of a fallen creature and presses the life out of it.

We file out of the building, and I cadge a cigarette from a woman farmer.

"Don't bother," she says to my offer of a dollar. "How else can we survive these things?"

The green lobby manages to drag out the approval for the bylaw until they can elect their candidates two years later. The first act of the new trustees is to throw out the vacation home bylaw and launch suits against anyone who rents their house to anyone at all. Thousands of man-hours and tens of thousands of tax dollars are thrown carelessly into the wind.

Chapter Four

Crony Conservation

When Bruce Chatwin wrote "an unsung land is a dead land," I'm pretty sure he didn't mean the song of the conservation bureaucrat. After five months of waiting for the trust to accept my application "into the process," my fretful phone calls and e-mails led Brent to deliver a flowchart, which lays out my future.

I shake it out of its tube, tip it onto the floor, unfurl it, and place paperweights at each corner. It measures three feet by five feet. It is titled Process for Subdivision by Way of a Density Transfer. The boxes are color-coded, with a key:

Yellow: Applicant
Pink: End or Termination of Process
Red: Land Office Registrar
Light Green: Ministry of Transportation
Blue: Local Trust Committee
Purple: Islands Trust Planning Staff
Mauve: Advisory Planning Committee
Green: Process Complete

At thirteen critical junctures amid the fifty-three steps, a pink box reads, "Application is rejected and project terminated." One box reads, "Applicant to engage a certified biologist to complete a baseline inventory of the donor area *and* check receiving property for specific biological areas that require protection." In other words, I will have to pay for not one, but two ecological assessments: $10,000.

To divide one twenty-eight-acre property in half, with two houses on that twenty-eight acres instead of one, I must pass muster with dozens of conservation bureaucrats, squeak through the meeting processes of two citizen committees, endure a public information meeting and a public hearing (each liable to excite deafening negative fervor), and proceed through four drafts of a bylaw (more fervor), each draft threatening failure. The trust then shops out the application to seven different departments from the ministries of Highways, Health, Community and Rural Development, Environment, and Aboriginal Affairs and the federal Department of Fisheries and Oceans. It looks to me as though the map represents hundreds of hours of work for bureaucrats and that my considerable fees—because at each juncture, there are fees—will cover barely a tenth of the cost. I am fully awake to the fact that members of the environmental movement on the island will bend their considerable energies to defeat the application at each of the fifty-three steps, thereby adding to my legal and consultant fees, the interest I'll pay on the loan funding this operation, the hours spent by bureaucrats paid by the taxpayer, and my misery index. And if this is what I'm facing with my modest plan, what fiend-ridden hell of a process map must face any businessman or businesswoman?

And so it begins. The ecological assessors turn up and disappear down into the forest with Brent. The trust is pioneering a covenant* program; if I promise to preserve in perpetuity part of

* Also called a conservation easement. It means I become a junior partner in the ownership of that part of my land which is covenanted. The senior partner, the Islands Trust, decides what I can do on that portion of land, if anything. In my case, it is a "no-touch" covenant, which means I cannot even "disturb vegetation." Fine until the forest, clogged with dead brush, goes up in flames one night and destroys the seven houses backing onto the ravine. The trust will not bear responsibility for those losses.

my land that the trust wants, I will get points and a perpetual property-tax deduction. So I attend the public meeting on this new program. Ten days later, a sylph named Kate Emmings arrives, and we trudge down into the forest. I know what she's thinking as we head across a narrow horizontal shelf toward the ravine: "So what, so what, boring, so what?"

"Oh, my, God," Kate says as the ravine opens up below us.

It is a bit like finding an enchanted realm. The assessors see the ravine as a natural save, and so does Kate Emmings.

Two hurdles cleared. Fifty-one to go.

I herd the dogs into the car and go to my mother's. She lives in the nearby city of Victoria, not far from the Islands Trust head office, where I purchase copies of every single one of their establishing documents. I shut myself in her spare room and read the following, for five eighteen-hour days:

1. Islands Trust Policy Manual
2. Islands Trust Fund Plan
3. Sensitive Ecosystems Inventory, East Vancouver Island and Gulf Islands, volumes 1 and 2
4. Islands Trust Legislative Monitoring Chart
5. Salt Spring Island Official Community Plan (OCP)
6. Salt Spring Trust Area Land Use Bylaw No. 355 (LUB)

Each is several hundred pages long, except the first document, the policy manual, which is over a thousand pages long. All this to manage a group of islands whose population of 25,000 is scattered over an area of 160,000 acres, a population, moreover, that is decidedly *not* growing.

I emerge, bleary-eyed and hysterical, having written my own twelve-page document called "How This Conforms," each of its forty-two points describing in detail how my density transfer and subdivision proposal conforms with the trust's policies and strategies. I point out that my project encourages and even demonstrates the efficacy of the trust's future legislative goals and goes far, far beyond the conservation goals of the Islands Trust Fund, the property acquisition arm.

Every point is accompanied by detailed citations and page numbers from each of the various manuals I've read. I'm reasoning that if those bureaucrats have to do this work themselves, it's going to take longer, right? So I'll do it for them! I send my document off to Brent for fact-checking, and he is enthusiastic, except that he spots the sarcasm I've buried in it by repeating certain inane phrases until even the very sleepy would realize they are being mocked. I cut the ridicule.

Excruciating delays are part of the process, I am told, so I descend into a fug. While marooned on my couch, I started wondering from which dank hell all this well-organized, if strikingly repetitive, verbiage originated. With only ten thousand residents, Salt Spring has more regulations than the province's capital city, Victoria, with almost four hundred thousand. So after I lost interest in *The Young and the Restless*, my sitting room became the repository of the founding documents of our current age. For the next few months, I lived in the midst of a paper storm.

According to the old-timers on the island, the Official Community Plan was written by self-selected citizens who formed committees, managing to shut out those considered insufficiently green. Members of these new committees then found an issue around which to organize and polarize opinion, a sewer for the only genuine village on the island, Ganges. The newcomers didn't want a sewer system, despite the fact that effluent was being released right into the harbor. A sewer would encourage development. Development was bad. It took two years, but by the end of the fight, those on one side would cross the street when someone from the other side approached. The battle was so bitter, angry, and slanderous that by the end of it, men and women whose families had lived on the island through several generations or who ran the businesses and stores—those, in other words, who would ordinarily sit on committees and town councils—decided to back away and leave the activists to it. It was local idiocy, they reasoned, not important, and they pressed the mute button. This was a pattern followed in tens of thousands of rural communities all over the world in the 1970s and '80s.

That decision played right into the hands of the newcomers, because while many of them looked and behaved like flakes, their ideas were anything but. The philosophical and scientific underpinnings of the trust's various policies did not come from a bunch of random hippies and draft dodgers trying to make their community groovy and earth centered. Quite the contrary. An Official Community Plan from the towns around the Great Barrier Reef, say, or Eureka, California, would tee up nicely with Salt Spring's. That is because every rural county claimed by the movement takes its cues, rationale, and techniques for "building capacity" and training from ICLEI, one of the UN's Agenda 21 progeny.* ICLEI is essentially an international lobbying and activist organization for local "sustainability."[1] That word is so overused that it has ceased to hold meaning, so let's call sustainability what Agenda 21 calls it: "changing consumption patterns."

The International Council for Local Environmental Initiatives, headquartered in Bonn, Germany, walks citizen committees bent on preserving "natural values" through writing the plans, negotiating those plans through senior bureaucracies, and pro-

* ICLEI covers its tracks in some rural areas, knowing how unpopular its policies are to traditional rural dwellers. Plans come from a sister organization.

moting them to the citizens of those towns and counties. More than six hundred towns in the United States conform to ICLEI's rules. Rather than being the supposed bottom-up approach that the promoters assert they are, the plans are anything but. Sustainable village or county plans are always top-down, whether the documents are bought from the American Planning Association or from someone from the Sonoran Institute—the two hot nodes for New Ruralist planning—or directly prepared in concert with ICLEI and the UN's Agenda 21 goals. Before you head for the door, which is what I did when people first brought up Agenda 21, we have to acknowledge that any organization that partners with the UN—any foundation or NGO, for example—is required to advance the UN's agenda. And Agenda 21 is the UN's social, environmental, and economic plan for the century preceding 2100. It is a soft policy; however, while there are no strict requirements and no annual benchmarks to meet, hewing to its direction is required. Penalties? Not being invited to posh resorts for endless rounds of conferences, not getting astonishingly well-paid contract employment researching and writing proliferations of convoluted verbiage, not being published, being denied tenure, and being shut out of the lucrative profession of sky-is-falling biology. De-development is Agenda 21's overarching goal, for how else are we to level the playing field, stop the world from warming, and halt the headlong rush to extinction? Agenda 21 is not a conspiracy; it's an action plan. The agenda and its various and many exegeses serve like the *Book of Common Prayer* for nonchurchgoing Episcopalians who still call themselves Christians. Accepted, read once if at all, Agenda 21's foundational documents are very, very boring and equally long-winded, but the ideas are so insidious and ever present that they form almost everyone's worldview. Nor is it toothless; since the Rio conference in 1992, most development aid from the West is funneled to developing countries through the UN's program for sustainable development, a key perhaps to the eternal conundrum: why aren't some developing countries developing?

Devising the remarkably fuzzy yet fiendishly complex metrics of environmental protection was left to a nonprofit multinational, the Nature Conservancy (TNC). The tenth-largest NGO in the world, the Nature Conservancy has advanced the rationale for land-use change using the science of conservation biology together with a plan to achieve what conservation biology demands: "natural ecosystem balance." Thousands of other nonprofits and foundations pitch in, but TNC is the biggest of the big dogs, the mythic wolf-king of the forest primeval, and so leads the charge.

The Nature Conservancy is headquartered in a $28 million, eight-story office building in Arlington, Virginia, but has more than 500 satellite offices in North American cities and towns, as well as in thirty other countries. From those offices, staffed by 3,200 squeaky-clean, if not angelic, give-backers, TNC from 2000 to 2008 spent $7,718,140,611[*] prosecuting its agenda, and with that money, threw a web of control over much of the planet. In late 2008, before the crash depreciated rural land values and portfolios, TNC's assets were routinely cited as hovering around $6 billion. That money meant that TNC controlled more than 100 million acres of the world's most biologically diverse lands. TNC has a reputation most corporations would die for; every year, it is chosen as one of the most trusted NGOs in the world.

But let's frame this outstanding achievement the way country people frame it. TNC has locked down 100 million acres,[†] containing trillions of dollars' worth of precious natural resources, which it now not only controls but has removed from the tax base,

* IRS 990s for 2000 to 2008.

† Not counting the acres it has acquired and sold on to the government to be included in wilderness.

ensuring that not even government will be able to "profit" by that land. Little wonder that some people call it the world's largest real estate cartel. And despite being cloaked in virtue, TNC has, as detailed by the 2003 *Washington Post* series on Big Green, logged, drilled for oil in protected species habitats, sold conserved land at a discount to big donors, and developed half of a "rare open sand plain" by building Gatsbyesque homes. Yet no one in the superculture, liberal or conservative, despite this extraordinary heist, questions its motives in any systematic way.

In 1974, with money from the Pew, Mellon, MacArthur, and Hewlett foundations, the organization began a tally of the world's biological wealth. Matching grants from governments[2] eventually totaled in the hundreds of millions, with which the nonprofit created something first called the Heritage Network. That name was judged too fusty, one imagines, because it was replaced by "The Network," which must have been dismissed for its Orwellian tang because in 2005 the data bank became known as NatureServe.org (modest, unassuming, cute—the opposite, in fact, of what it is). NatureServe has mapped more than one-third of the planet. The *Wall Street Journal* reported that the database is so fine-grained that it records the precise locations of individual eagle nests and clumps of endangered plants,[3] which I suspect is a threat nicely disguised as propaganda. But if a private tract with valuable assets is listed by NatureServe, government attention is focused on the property and it is generally accepted as being on the list for acquisition, or greenlining, thus destroying its market value, crashing its tax value, and distorting its price. If you own property in the Western Hemisphere, the Nature Conservancy knows what's on it, or thinks it does. And if it's a nice older-growth forest with a magnificent ravine, for instance, you can be sure that someone associated with TNC has her eye on it.

Most senior government officials in other countries, rather than spending money to duplicate the Conservancy's work, use NatureServe data and policy to make their own land-use decisions.* Wherever you live, you probably have something called the Conservation Data Center (or some similar name) set up for your regional government by the Nature Conservancy or one of its partners. Its data is what your land-use bureaucrats use to make decisions on just about everything that happens in your area. Has the data from your Conservation Data Center ever been groundtruthed? Have independent surveyors ventured out into the actual forest, range, field, or mountain to check that data? Almost certainly not; fieldwork is prohibitively expensive. Your data was largely modeled, desktopped. If any fieldwork was undertaken, it was probably done by undergrad biology students thrashing around in the bush trying to get their bearings[4] but nevertheless seized with the fervor to protect anything they found—and not merely because claiming species endangerment ensured them of future employment. TNC gives young biologists messianic goals, and who among us does not want to save the world and make a good living at the same time?

In 1995, the Conservancy launched Conservation by Design. Using "a cutting-edge approach"—called the ecoregional approach, which identified biogeoclimatic zones—it set priorities on acquisition of valuable land. Employing a collection of techniques, many developed for the control of private property in Adirondack Park: eminent domain, forced purchase, conservation easements, acquisition of land parcels in a pattern designed to bankrupt unwilling sellers, promises of new jobs in ecotour-

* The Bureau of Land Management, the Forest Service, the Environmental Protection Agency, and particularly the Fish and Wildlife Service, with its secretive Land Acquisition Priority System, and the National Biological Information Infrastructure (NBII) are provided "increased, integrated access to selected data from TNC's central databases, including their rich geospatial data sets."

ism to replace jobs lost by resource workers, and when possible, clearing entire counties of resource workers, thus turning lands back to wilderness, the organization began its relentless advance on natural beauty spots everywhere.[5] Hardly anyone thought this was spooky. In fact, quite the opposite.

TNC became masterful at corralling the famous from the media, Wall Street, Hollywood, CEOs of the biggest banks and corporations, senior partners at the most august law firms, famous politicians and generals, and prominent scientists and academics into its fold, creating the illusion of buy-in at every level of the culture. The strategy served to make them untouchable, almost impossible to criticize. Indeed, how could anyone criticize the NGO? It was saving the very essence of life, its biodiversity. It was doing God's work. Everyone agreed. Therefore, at one time or another, Norman Schwarzkopf, Janet Reno, Gerald Levin, Joanne Woodward, Paul Newman, Diane Sawyer, Jessica Alba, Robin Wright Penn, of course, and Alec Baldwin supported the organization either by serving on one of its several boards or helping promote a Nature Conservancy campaign. Many other celebrities made formal appearances at parties, notably, as Timothy Findley reported, at one Central Park celebration where the honorary chairs and masters of ceremonies for the exclusive $750-a-plate event included Peter Jennings, Dan Rather, Paula Zahn, Charlie Rose, Mike Wallace, and Charles Osgood, at the time the cream of elite media. Prominent bankers and politicians were equally happy to serve on TNC's national and regional boards. By the late 1990s, the Nature Conservancy *was* the establishment.

It wasn't until 2003 that TNC's virtuecrats suffered a takedown by any mainstream medium. Based on the stellar reporting of Findley in *Range* magazine earlier in the year, the *Washington Post* ran a twenty-two-piece series that triggered a near-panic in the Conservancy's offices, which triggered a full-bore sleazefest of misinformation, crisis management, and counterpropaganda and finally a piece of hagiography full of nature porn, faux modesty, confused fearmongering, and lovingly described Potemkin villages, which was called *Nature's Keepers: The Remarkable Story of How the Nature Conservancy Became the Largest Environmental Organization in the World.*

Senator Charles Grassley began a Senate investigation, which observed that the NGO's financial reporting was so murky and its arrangements so convoluted that no one could really figure out what was going on, except that there were many unusual transactions that should have been more thoroughly and accurately disclosed by TNC in its IRS 990s. For instance, the Conservancy, after shearing off development rights, routinely made below-market land sales to donors and insiders, who then donated the difference in the price to TNC. One thing was for sure: transparent it was not, nor perhaps, was it charity. "The Nature Conservancy (TNC) must make a firm commitment to ending recent abuses that have called into question its compliance with federal laws bestowing economic favor on charities," concluded the U.S. Senate Finance Committee report issued on June 7, 2005.

Much like many Clinton-era private-public partnerships, like Fannie Mae and Freddie Mac, the Nature Conservancy swiftly became practiced at increasing its wealth and scope by using new financial mechanisms. Pacific Gas and Electric, Dow, GM, BP, Exxon, Southern Electric—1,900 corporate associates paid $25,000 a year for annual conferences wherein they learned how to use tax breaks for conservation, as well as, of course, other ways they could "work together to preserve biodiversity." Some were notorious polluters and continued to be so according to the EPA and the Environmental Defense Fund. More corporate than many multinationals, TNC knew every trick. As the *Post* said, "The organization has many of the trappings of a Fortune 500 company: global reach, consumer focus groups, meetings with world leaders, sophisticated marketing and cost-benefit analysis applied to conservation." The NGO also developed sophisticated arrangements with federal government departments by partnering with the EPA, the Fish and Wildlife Service, the National Park Service, the Forest Service, and even, as we will see, the U.S. Army in land acquisition. Some owners were loath to sell their land to the government; TNC typically did not tell those sellers that the ultimate fate of their land was that it be public and, often, mothballed.

Are you surprised that former Secretary of the Treasury and Goldman Sachs CEO Henry Paulson was chair of the board of governors and on the board for the better part of a decade? During Paulson's tenure, representatives of the American Electric and Power Company, at the time called the worst air polluter in the United States, and Georgia-Pacific, the largest paper producer in the United States, sat on the board. Arrangements with companies of which TNC board members were principals were frequent, as the Senate Finance Committee's staff report on the Nature Conservancy stated. And while the conservancy pointed out that there was nothing illegal about these transactions, TNC had been worried about these associations for a while. As the *Post* reporters pointed out, in 2001, it had commissioned an internal study that tested participants on their reactions to TNC's hypothetical associations with BP, Walmart, Amoco, and so on. Most considered such associations as negatives. The study warned TNC against too much corporate involvement, saying that it would tarnish TNC's brand. And in fact, David Morine, a longtime TNC employee who had pioneered the corporate tie program, said after he left the organization that creating it was the "biggest mistake of my life. . . . Those corporate executives are carnivorous. You bring them in, and they just take over."

Greenwashing remains a problem in the environmental movement. TNC and its thousand cohorts will greenwash a project or product if it receives a benefit elsewhere—a donation, land, goods and services in kind. While there is nothing illegal about greenwashing, it remains an unseemly fact of the modern movement. Full disclosure would tarnish the brand. But more important, greenwashing of a corporation can obscure real problems that require immediate solutions. The Senate committee reported fifteen other transactions between TNC and its board members, all of which raised red flags, some of which should have been reported on TNC's IRS 990s and were not. As the *Post* pointed out, many of the corporations whose principals sat on TNC's board had paid millions in environmental fines, but in one year alone they had donated $225 million to the TNC. It didn't look good. Nor, reported the *Post*, were special deals limited to board members. The *Post* charged that the president and CEO at the time, Steve McCormick, had received a significantly good deal on a jumbo house loan of $1.55 million; the staff Senate report affirmed this finding and stated that it gave the appearance of self-dealing since it did not form part of McCormick's compensation package.[6] McCormick paid the loan in full after the *Washington Post* piece appeared.

But it was TNC's "conservation buyer" program that gave the Senate committee and the *Post* the most pause. It turned out that while some of TNC's "saved" land is actually set aside, a substantial portion of it has been developed industrially—oil wells, golf courses, condos—and it appeared that conservation buyer programs mostly benefited the very wealthy.

Conservation buyer programs work like this. A TNC member or staff identifies a desirable piece of property. TNC purchases that property, often at a discount because it is a land preservation charity, and the owner of the property receives a tax write-down from the appraised value of the land. TNC then strips the land of development rights, that is, the right to cover that land with industrial waste, trailers, tract homes, strip malls, and so on. It then sells the land on to a private owner at a discount. The new owner can build a house, cabin, stables, septic system, dock, etc., but (generally) can't develop the land as anything but an estate. As the *Washington Post* pointed out and the Senate Finance Committee affirmed, all too often that new owner was a board member or member of TNC. The new owner then donates the difference in the price of the land, if developable, between its maximum and the price he paid, to TNC, receiving a tax write-down of that amount. Everybody wins, except the taxpayer. The original seller wins a tax write-down; TNC wins the land at a reduced rate, sells the land on, and then receives a further gift. Was the land worth what TNC and the seller decided it was worth? Appraisers, noted the Senate Finance Committee, had been known to inflate the value of property. None of this, I stress, was illegal. It was just dubious, and unseemly. As the Senate committee asked, was it consistent with TNC's tax-exempt status and federal tax and information reporting requirements? In many cases, the Senate Finance Committee decided, it was not.

But for a while, it was simply splendid for the good and the great. According to the *Post*, the biggest and best-known easements have been linked to the nation's richest individuals, like Ted Turner, David Letterman, the Rockefellers, and the DuPonts. Basically, TNC is acting as agent for the wealthiest among us, acquiring enormous tracts of land, using $25 donations from its 1.3-million-strong membership and $100 million in annual government money, and then *selling that land at a discount* to the very rich, who in effect receive a substantial tax discount as well as an extremely beautiful place in which to establish a country estate. That doesn't count the moral bump they get in their self-esteem and the status of being included among the most privileged group of people who ever lived. Conservation today acts much like indulgences sold by the Catholic Church, except that instead of paradise, the donor gets perpetual tax deductions. As the *Post* reported, one lucky woman's father, a Conservancy state trustee, brokered the deal with TNC for her horse farm in Kentucky. She pointed out that she had built a six-bedroom house and a twenty-stall horse barn and grazed cattle on her land. The easement had not affected her land use in any way; in fact she might have bought the tract and used it in the same way without the Conservancy's involvement—and without the gift from U.S. taxpayers. The easement continues to save her money. She said the county assessor values her 146 acres and the new six-bedroom house at $150,000. "It was so low, I laughed," she said.[7]

After the *Washington Post* series, and during the IRS investigation concurrent with the Senate Finance Committee investigation, the organization admitted it had been operating "outside our own headlights." Before either the IRS or the Senate had reached their conclusion, TNC, in a board meeting on June 13, 2003, formally stopped all drilling for oil and gas on Conservancy properties and all for-profit ventures, and adopted internal policies to govern cause-related marketing, insider conservation buyer schemes, and special loans to employees.

Which didn't stop Senator Grassley from weighing in: "I'm troubled enough when I see the words 'tax shelter' appearing in tax planning documents of for-profit corporations," Senator Grassley said, describing a Nature Conservancy document. "When I see 'tax shelter' being used in documents of charities, we ought to be really worried."[8]

I, too, with my ever so modest covenant on my ravine, get a tax deduction that is worth, over the twenty years I hope to live on this property, between $10,000 and $15,000. Analysts rightly doubt the value of those write-offs. While my covenant's value is set by the trust and the tax assessor, the Government Accountability Office (GAO) estimated that in general, easement or covenant owners overclaim the value of their easement by 220 percent.[9] As Joe Stephens and David B. Ottaway reported in the *Post*, "In the Great Smoky Mountains near Asheville, N.C., investors two years ago bought 4,400 acres, placed an easement on 3,000 acres and then began developing 350 home sites and an 18-hole golf course on the remaining property. A master plan for the development, called the Balsam Mountain Preserve, shows that the easement area is broken up by the fairways and home sites, which spot the land like mushrooms on a pizza."

Investors paid about $10 million for the land and shared in a tax write-off of a gob-smacking $20 million, according to James A. Anthony, a partner in the South Carolina development firm of Chaffin Light Associates. The deduction was based in part on an appraiser's assessment of how much the land would have been worth if they had filled the acreage with 1,400 homes, Anthony said. As the *Post* reported:

> *Mike Kahn, a Florida business consultant and former golf pro, advises celebrities and sports stars how they can save millions in taxes: Buy a golf course and prohibit building on the fairways.*
>
> *"You make virtually risk-free easy money," Kahn's Web site says. He explained in one Internet posting how an investor paid $2.4 million for a golf course and reaped $4.8 million "in pure tax savings." Kahn will not identify the buyer but describes him as one of many who made big money—and got to keep the golf course as well.*
>
> *"People who do it generally keep it quiet," he explained in an interview. "It sounds like a money grab."*

Such tax bonanzas have become a little-noticed byproduct of the maturing environmental movement, which increasingly entwines preservation of land with preservation of wealth.

While the Balsam Mountain development didn't have its tax exemption pulled, it and other suspiciously rich golf-course easement deals triggered a full-bore investigation by Steven T. Miller, the commissioner of the tax-exempt division of the IRS, who said in the *Chronicle of Philanthropy* in 2005, "Potential valuation problems with conservation easements have led the revenue service to start 240 examinations of donors who have taken an open-space easement deduction. The IRS is considering examinations of another 100 donors who have participated in easement transactions [and] is looking into the easement practices of many charities. Some donors appear to be taking a large deduction for a conservation easement that is not used for a charitable purpose. One popular easement abuse involves golf-course developers."

Hello, just how fabulous was this while it lasted ? And while this is all interesting to us peons because we can see the most powerful and famous people in the United States systematically looting the wealth of the country *and* ruthlessly deploying seemingly witless celebrities like Tom Hanks and all the usual suspects, the real damage, the structural damage, happens where no one is looking. After the *Post* series, TNC largely stopped its insider deals but stepped up its activities outside the celebrity-hedge-fund–Hank Paulson biogeoclimatic zone considerably.

While the *Post* didn't get too deeply into TNC's operations in working country, *Range* magazine did not pull any punches. In his 2003 "Nature's Landlord" takeout, Tim Findley calls TNC "the world's largest real-estate cartel or worse." Indeed, few people in working country display the supine position of the average city dweller when it comes to TNC; in the heartland, TNC is hated and feared, its name uttered in whispers, as if it were the medieval devil. As a result, TNC operates through a proliferation of "partner" land trusts, conservancies, and operatives. TNC's sending polite, fresh-faced kids into the middle of nowhere to start local actions for waterbirds or watersheds or ancient forests was the trigger that started the landslide collapse in rural America.

Nor did the Senate, IRS, or *Washington Post* investigation slow down the money cascade. From 2002 to 2008, TNC received $589,303,090 in government grants and $2,900,618,368 from the government for land sales, and its total receipts for a five-year period were over $13,432,536,073.[10]

"Listen for the frailty."

That was the order that John Sawhill, past president of New York University, then head of the Nature Conservancy, gave to thousands of youthful operatives in the 1990s. Sent into rural communities, they were to find a way to acquire an asset—a river, a wetland, a fertile valley, or a forest—that the Conservancy wanted. Join clubs, volunteer, help out, be a neighbor, play with their kids.[11] And listen for the frailty.

But many times, that valley or forest wasn't all they wanted. When Harry Reid wanted the water of the Lahontan Valley for Las Vegas, the only way he could get it was through John Sawhill, who sent Graham Chisholm, now the executive director of Audubon California, to buy those rights. Chisholm's master's thesis was on the development of the Green Party in Germany, and in the five years he spent romancing the citizens of the Lahontan, Chisholm demonstrated the iron fist cloaked in Robin Hood suede of "the largest environmental organization in the world."

Chisholm arrived just after the 101st Congress, in the last act of its last session, delivered to Harry Reid something he'd been after for a very long time. The Settlement Act gave Reid the right to "settle" the dispute between California and Nevada about the water splitting off the west and south flanks of the Sierra Nevada. The Lahontan Valley was the site of Teddy Roosevelt's first western reclamation project in 1906, when he promised to "make the desert bloom." Ranchers and farmers had vested water rights in the valley that went even further back than 1905. But as soon as the act was signed, Lahontan Valley farmers were beset by fresh regulatory demands and lawsuits.[12] Since they had endured seven years of drought, the regulations meant that many families were teetering on the edge of survival—which is exactly where Reid and TNC wanted them.

Here's the thing about country people. They live by the rhythms of nature; some years some crops flourish and others don't. There may be long periods of drought or too much rain. Or too much cold. Or heat. You have to be able to turn on a dime and grab opportunities before they vanish. You do not have time to purchase a permit, hire registered environmental professionals to do a study, and then wait for months, if not years, for bureaucrats a thousand miles away to give you permission. You start a business, make some money, end it in eight months, and start another. You sell a hunk of land one year and buy it back the next or ten years later. You shear off an acreage for a child who wants to farm or ranch or to build a house for an aging parent. You build shacks for seasonal workers that are in use for ten years, then tear them down or use them for WWOOF (Willing Workers on Organic Farms) or agritourists or a straightforward motel for resource workers because a mine or gas field has been opened in the area. You make your living from a half-dozen different sources, and pretty much every year, the sources change. Regulation that controls everything from the height and footprint of a barn, creek setbacks, and the number of outbuildings to the number of people you can employ, how you must pay them, the depth and flow of your irrigation system,[13] and the dust and noise you can raise make life pretty close to unmanageable.

But for Chisholm and the Lahontan Valley, the first years were a honeymoon. Thrilled to have a young, educated kid move *into* the valley, especially one with such starry connections, residents treated Chisholm like a celebrity. He did not disappoint; according to Findley, he joined everything. He volunteered with the Ag Center's committee testing local wells in the drought. He organized a new group to defend local water rights. He was at the meetings of the irrigation district and the county commission. He was at the melon festival and the county fair. This kind of behavior is duplicated in all 3,300 counties in the United States, where agents of TNC or any one of a proliferation of foundations or ENGOs work to determine the future of the community. On my islands, activists from Montana, Los Angeles, New Jersey, and a half-dozen other states moved here and now chair a dozen political action committees, all of which produce hysterical reports based on "science" about our collapsing ecosystems. Our Salt Spring activists are in their creaky sixties; the island is more like a retirement camp for successful environmental activists than a place where the young can earn their stripes. But in any region, you can tell who they are by the causes they espouse and the issues they press: water, first and always, then affordable housing, the saving of whatever natural feature dominates the region, and of course the sleeper, climate mitigation. Well-educated, well-spoken, and always talking about "the community," before a marina or hotel or an organic fair-trade coffee-roasting facility can be built they insist on "studies" that will cost the public purse in excess of $100,000 and take five years. I have not found, in a decade of searching, one authentically rural region to which they have not brought decline.

Or disaster. By the time Graham Chisholm left the Lahontan Valley, it was already trending toward dust and tumbleweed, the houses abandoned and useless without a working ranch surrounding them. And the people found they'd lost their purpose as well; without the connection to the land they'd had so long, they didn't really want to live. As Timothy Findley began his story on the fate of the irrigators at Harry Reid's hands, one rancher, the day after signing all the papers that transferred the rights to his ranch to the Nature Conservancy, went out on his back porch, a check in his pocket that meant he could live anywhere he wanted, and shot himself. Perhaps he had cancer; perhaps his wife had just left him; there was no note. The salient facts remain. He had just received more money than he'd ever had or even dreamed of, but he didn't want to live. Wherever you go in ranching country, similar stories proliferate. Mostly, people work so hard at staying, at conforming to the mitigations, regulations, hearings, lawsuits[14] foisted upon them, they fall ill.

It all sounded so good. Going in, Chisholm and TNC claimed they merely wanted to "restore" 25,000 acres of wetland on the Carson River, ignoring the fact that the Lahontan farmers had kept it as a duck preserve for decades. But it was win-win. Chisholm's offer meant they could keep the marsh and get some badly needed money. Given the drought and regulatory compliance costs, community leaders admitted that there had to be some additional sacrifice going forward. Marginal lands would have to be sold. Chisholm offered to buy them for the Nature Conservancy. The head of the irrigation district was the first, then a leading family, and so it went. Children sold their heritage, more interested in the money than a lifetime of brutally hard work and uncertainty. But the lands weren't bought at their appraised or fair value, because only the water rights were being sold. It seemed fair enough to the desperate. As is typical in this game, once the water rights were gone, anyone left was forced to sell at a considerable discount; TNC had effectively reduced the value of Lahontan Valley land to virtually nil. Said the last holdout, seventy-five-year-old cowboy poet Georgie Sicking, "I had no choice, really. They bought everything around me, including the irrigation ditch."[15]

But Chisholm got his marsh "save" and a promotion to director of the wealthiest chapter of the Nature Conservancy in California. He was praised and celebrated and today is acclaimed for "saving" countless "treasures," with no mention of the ruination in his wake. But by the time of his promotion, the affection for Chisholm in the Lahontan Valley community was long dead, especially when they discovered that Chisholm had been operating under a memorandum of agreement between TNC and the Fish and Wildlife Service to buy the water rights and shift them to the government, leaving, as Tim Findley put it, "ghost farms and pale, lifeless fields of laser-leveled ground unable even to support the growth of weeds."[16]

Needless to say, Harry Reid got his water. And rather than the abundant nature touted in its nature porn, the desertified Lahontan Valley is a more typical Nature Conservancy gift to the future.

Most of us blithely ignore this level of dubious conduct and ruination, and the reason is more complex than general lassitude or narcissism. We are obeying a deeper psychology, one that plays right into the hands of ENGOs and the foundations that direct their actions.

We're rich enough to hate what we've made. So much of our built environment is blight, and our lives in those strip malls and featureless modernist towers are so sheared of beauty and mystery that every trip to the country or seaside corrupts us and makes us hate our built landscapes even more. James Howard Kunstler, before he went bog-crazy with *The Long Emergency*, a truly demented work hailing the crash of Western civilization, nailed our discontent in his 1994 breakout bestseller, *The Geography of Nowhere*, and later in *Home from Nowhere*.

We all want what we think our grandparents had, a sense of place, of belonging, a home that we don't want to leave. Grace, clean air and water, the beauty of trees, the glory of mountains and meadows and watercourses—that's our hunger. Kunstler calls this a "more spiritually gratifying way of life . . . the idea of a true community organized at the human scale, along with a feeling of secure remoteness from the so-called modern world and all its terrors of giantism and discontinuity." Never mind that this loss has been largely brought to us by post–World War II, top-down authoritarian planning; what we need, say Kunstler and his New Ruralist, smart-growth, New Urbanist fellow travelers, is yet more planning, this time to machine those "true" communities into being. Hence Celebration, Florida, or Prince Charles's favorite town, Seaside—spooky Disneyesque reconstructions of village life circa 1910.

Nevertheless, we are proving our distaste for our rigid cityscapes with our feet. The U.S. Department of Agriculture has noted for the past twenty years a steadily increasing trickle of urbanites moving to the American heartland. After 2001, that trickle became a flood, and in 2009 the department released a report titled *Baby Boom Migration and Its Impact on Rural America*. Based on hard numbers, demographers charted that an astonishing 30 percent population increase will occur in rural America from 1990 to 2020. These migrants are not traditional goods producers; if they farm or ranch at all, they will do it as a hobby or for the tax deduction. Nor are they projected to be solely vigorous retirees; in fact, by 2009, everyone from the *Wall Street Journal* to

Nielsen PRIZM's marketing categories noted the flood of youngish families called "New Homesteaders" looking for the kind of country life that perhaps only ever existed in novels or television series but that, nonetheless, they are determined to reproduce.

We fund ENGOs and lie down for Agenda 21's secretive initiatives because those organizations have identified city people's hunger. They promise that if we give them money and power, they will bring us that connection. Urbanites assuage their hunger and feed their country dreams by funding ENGOs, oblivious of how that money is used. Then, just like me, they move to the country and find themselves enmeshed in limitations, the measure of which I am just beginning to calculate.

Philosopher of science and writer Alston Chase, the most trenchant, credentialed critic of the environmental movement, sees it as an aesthetic and philosophical movement that, armed with bad science and worse politics, metastasized into an unstoppable and now broadly destructive force. Chase traces the science, which "unleash[ed] a tidal wave of planning, designed to analyze and control every square inch of American real estate."

I'm beginning to think that the planning to which I am trying to conform in order to save my life on my land is not designed to help me, the community, or the land but to create that perfect world of empty space and mystery. It wants to prevent me from doing what I want. Am I meant, like thousands of failed applicants, to bang my head against a wall, spend all my money, exhaust my patience and determination, sell my land, and leave? Surely not. But this is what my reason—given the experience of others—tells me. Some days I think I'm in what a friend calls the Chinese Cultural Reeducation Camp, where weekly I am made to swear, to one green gatekeeper or another, allegiance to all things green and all people who advance green, no thought or questions allowed. If I demonstrate perfect conformity with the trust's policy manual, the island's Official Community Plan, and its Land Use Plan, perhaps I will be allowed to stay.

Today I can apply for a permit that, *under law*, I am allowed, and immediately the entire bureaucracy and what is now called civil society, consisting of activist fellow citizens who disagree with the law, will bend themselves to thwarting me, no matter the effect on my bank account or the number of tax dollars they will spend thwarting me. Their actions are antiprosperity (or "sustainability") or a forced "changing of consumption patterns." If I lose, as I am meant to, my consumption patterns will very definitely change; I won't be able to consume much at all. But the money spent by activists and the bureaucracy on thwarting me will not only not create any wealth for me but will drain the public purse in a small way, small enough to not be noticed. The only people who profit are the bureaucrats and lawyers who negotiate my doom. And of course, the movement has another scalp for its trophy wall. Or rather, another "save."

It's All About the Salmon?

Brent promises action, and sure enough, the next morning a breathless phone call summons me to the lower meadow. I leap from my desk, pull on a jacket and boots, and plunge into the forest, the dogs following. Sunlight streams bars of light through the trees, but the rain has been so persistent that mist unfurls in whooshing clouds from the depths of the ravine. We rush along the shrouded creek and down the last hill to the meadows. They are the very picture of classic ruined landscape, essentially six acres of invasive weed, though still beautiful to me. In the lower meadow, the orchard grass is high, concealing the disaster beneath. Both upper and lower meadow are ringed by maple, cottonwood, cedar, hemlock, and fir. The creek sounds white noise. The forest was logged thirty years ago, and the meadows were stripped of their topsoil. The lowest meadow, where the creeks lie, had been used as a gravel pit for a few years after that, until the pit silt fouled the water downstream. The only shrubs that grow in dirt stripped of nutrients are broom, thistle, stinging nettle, and blackberry—classic edge habitat, filled with below-stairs life: rodents, frogs, toads, newts, quail, fungi, and millions upon millions of insects.

The dogs start, then rush at the salmon lady, Kathy Reimer, who is already working. Steam rises from the creeks in a dense mist, refracting sunlight in a halo around her as she darts from streambed to creekside. She looks up and calls: "Look at those rocks. They're big. That's good, that's good for us!"

"Why?" I hurry over.

"Because of the flow . . . get down now!" The dogs are all over her.

"Big rocks are carried by fast water. Look at the depth of the streambed. It must be amazing down here after a storm." She has a high fluting voice, but I hear grit in it. I stand, looking into the creek, trying to see what she's seeing, until I realize she is striding toward the other creek.

We rush after her and trudge up a steep hill. I've never been this far into the forest. We are halted by a moss-covered wonderland.

"Oh, my," she says, and wheels back. She heads back to her car, and I stumble after her. She pulls out a map, struggles with it in the damp air, finally unfolding it on the hood of her car.

"See where they join? They are part of the headwaters of Fulford Creek, for sure!" She throws her arms up in the air and dances for a moment. The map rustles in the wind and starts to blow away. We both rush to grab it, and I find myself grinning at her, feeling something distinctly odd, I want to say carefree.

She folds up the map, shoves it in the backseat of her car, and charges back to the creek intersection. I follow, fascinated. Another set of incomprehensible terms flows out of her mouth.

"Who *are* you?" I finally ask.

She looks at me and grins, then laughs. She is who she is, a country woman: comfortable in her body, rumpled, tired, hair barely brushed, clothes well worn. Her small black eyes bounce with laughter as she cracks one joke after another until, in a sharp reversal of mood, she tells me she and her husband originally moved to Salt Spring after her baby daughter died from leukemia. The cancer, she believed, was triggered by toxic exposure in utero, while she was restoring a degraded site saturated with dumped chemicals.

"Do you think land has memory?" she asks. "Does it take revenge? In the valley I grew up in, way up in the interior, they raped the land, mined it, logged it to flat earth. Since then, the families in my valley have experienced one terrible disaster after another."

The subjects are changing too fast.

"My mother used to hit me, so I would escape into the bush out under the stars. I'd just stay out there, alone for days. And then I began to hear the music of the spheres."

"Really?" I say. I am dumbstruck, possibly bewitched, maybe because she has soothed my fear, when she says, in another sharp turn: "Of course, you should have your subdivision. Oh, yes, this is a natural." Relief floods my nerves. Kathy's casual assent, added to her easy assumption that this mess I own is restorable, means I am powerfully engaged.

The pace quietens now that she has made her discovery, and I begin to ask questions. I hardly need to. Used to property owners at sea, she reels off her credentials as we walk the creeks once more, while she pencils notes on a scrap of paper. She's spent the past twenty-five years knee-deep in the waters of the island, looking for salmon fry, making them places to shelter in a drought, clearing their waterways so they can come back to spawn. Every so often, she goes back to school to take a course or pass another certification in some aspect of riparian restoration. But her genius is folk; she sees another island than the one we mortals see, this one underground, with shale seams eight hundred feet deep, underground springs, and glacial-cold well water.

"At another time in history," she says, "I would have had to go underground myself or be declared a water witch and burned at the stake."

She has had considerable success; by building fifty spawning ponds, she's brought back salmon to twenty Salt Spring creeks. Today, at the bottom of my property, she projects a pond, not in the creek bed but beside it, the creek banks planted with red dogwood. The heritage orchard she thinks used to be here, reestablished, chocolate orchids nursed back, camas lilies popping through the wetland like flags, salmonberry bushes planted on berms that border the road to protect the extraordinary world, an almost precontact world that we will have restored. She'd be glad to teach me how to care for it; in fact, she has a nursery filled with native shrubs and trees from which I can draw.

I am six again, in the capable hands of an older girl.

If she has calculated correctly, the salmon will swim upstream from the ocean three miles away, through my neighbor's place across the road, through the culvert under the road, and upstream to my ravine. Therefore she will dig a pond at the juncture of the two creeks, leaving a berm to separate pond from creek, hoping the salmon will leap the berm and spawn in the pond. The pond will contain two deeper recesses for salmon fry to over-summer. Once the process is in play, she will start linking my restored watershed to others.

Restoration or conservation biology, though it is new science, wallowing in untested assertions, says that for the salmon enhancement project to work—really work, not just provide a talking point or virtuous gesture—the biota, the collection of organisms that used to live in this meadow, around these creeks, must be reestablished. That means removing broom and thistle. Scotch broom, I've discovered through grim experience, must be killed by hand, cutting it at the base every year just as it flowers, for three years until it gives up. Weed whacking or chainsawing causes it to multiply because, like the Scots themselves, the shrub *likes* adversity. For every bush you pull out by its roots, ten more start up, because now you've covered the waiting seeds with dirt. Then the fawn lily, camas lily, bogbean, marsh clover, Nootka rose, and salmonberry must be carefully introduced into the right places in the meadow and protected, which means fenced.

The rational voice shoulders its way through the dream Kathy presents. "Ten years' hard labor? A hundred thousand dollars?" An experienced country woman would recognize that I had just welcomed a charming interloper into my life, one whose dreams will allow her to annex my property, using the covenant and salmon project to supervise and restrict my activity, becoming for all intents and purposes the senior partner on my land. Let's use the word country people use: *taken.*

CHAPTER FIVE

Inch by Inch: Biocentric Command and Control

Edmund Burke's 1752 *Philosophical Enquiry into the Origin of Our Ideas of the Sublime and the Beautiful* for the first time placed awe, terror, and the solitude found in forests, chasms, wastes, and stormy seas on an equal footing with the neoclassical values of equilibrium and order, restfulness, smoothness, and rationality. Out here, astonishment, the loftiest state of the soul (to Burke's mind at least), is commonplace. Here we have taken nature as the face of God up a notch. We want to live right in it, preserve it, and model ourselves upon it. The rough, seemingly careless strokes of nature are perhaps the most powerful intoxicator of all.

I am no different from the fanatics and true believers who created this unreality, though I am not so elevated as to think of saving anything but myself. I need green as I need air and water; I wilt when away from a forest and fields, as if part of me goes into eclipse. I grew up in the country, on a three-acre plot of land behind which lay a forest, manicured a hundred years before for late-nineteenth-century promenaders. The once wide paths were overgrown, and the teahouse had collapsed, the brick pillars at its entrance pulled down and broken by the weight of ivy. I never saw anyone else in that forest. A train track cut through it, the lake was on the other side of the tracks, and I had been transplanted from a city at the age of six to this place where hours of freedom reigned. Well, that was everything that mattered. I had friends, of course, and countless activities, but when I think of my childhood, I think of myself, alone in that forest. By the time I was fourteen, school whisked me away, then university, then a career, marriage, and all the shiny gift boxes of a supposedly successful life. And then one day, all of a sudden, none of it mattered anymore. What mattered was coming home.

I had not anticipated that I would have to spend years defending the life I had built, but at some juncture, I stopped panicking and started enjoying it. I spent my days dashing back and forth across the deck from the main house to the study I had built, on bare feet in the snow or pelting rain, because I liked it. I had grown so resilient that one night, around midnight, I found myself in rubber boots and a dressing gown, ax in hand, in a blizzard so intense the wind was blowing the snow sideways, needles landing on my skin. I was splitting wood in anticipation of a power outage and shouting with exhilaration.

I had never been so healthy. And riding the storm of bureaucratic process was occasionally terrifying—my bills mounted precipitously, and every month there was another delay—but that was exhilarating, too, because the odds of making it were diminished.

"We have wrapped the world in ineffability," says Alston Chase. Of the battalions of writers on conservation and environmental protection, he is the only one—and I find this disturbing—who has done the necessary forensic research on how that protection has been effected. Chase, a philosopher of science trained at Oxford, Harvard, and Princeton, chronicled the results of that protection in two books: *Playing God in Yellowstone: The Destruction of America's First National Park* and *In a Dark Wood: The Fight over Forests and the Myths of Nature.*

After their three sons were fledged, Chase and his wife, Diana, also a professor, relinquished tenure and moved to a three-thousand-acre ranch in the Smith River canyon in Montana. The ranch was fifty-five miles from the nearest town, ten from the nearest neighbor, and thirty-five from the nearest county road. It had no telephone or electricity, and the only buildings were shacks. For almost ten years, they lived there with their animals, rusticating, writing, and running wilderness camps for kids in the summer. They kept a tiny house in Livingston, called the phone booth, because they used it mostly to make phone calls. It took a day to get there. But once Chase dug into *Playing God* and realized what it entailed, he and his wife sold their ranch to move closer to Yellowstone.

In both books, Chase picks apart the building of the science and the way it was used by the rising diaspora of activists. He chronicles, year by year, how that activism metastasized through the culture, forcing legislation, coming to dominate social and political life in the hinterlands.

After two decades of studying the evidence, interviewing the players, and following each trail to its end, he came to one unshakable conclusion: we got the science wrong. When I turned the last page of *In a Dark Wood* for the second time, I got into my car and drove to Montana.

The sky is a blue-gray haze lit from behind by the sun; it is so eerie and beautiful, it is as if I have been transported to another universe with a different color wheel. Eighteen inches of fresh snow sits on the roof of my car, and I must brush it off in such a way that I don't spend the next few hours sitting in wet clothing. Not as easy as it sounds, so by the time I have managed that, the sky has darkened to gray, and white flakes whirl and gather in velocity.

Sleet is driving into my windshield by the time I set the GPS and inch, ever so cautiously, onto I-90 to Livingston, thirty miles and an eternity of black ice over a high mountain pass. The road is virtually deserted, so it doesn't matter that it feels as if the white film of cataracts covers my eyes.

Livingston is a town of fewer than seven thousand and lacks the usual corporate blight on its access highway. Like all boutique country towns, it is beset by its own celebrities. Jeff Bridges and Margot Kidder roost here, as does Jim Harrison, author of *Legends of the Fall*. But mostly this is a town that services ranchers and the guide outfitters who ply Yellowstone, on the slopes of which Livingston rests. The town is easy to navigate, with a sweet, small inner residential area, a serviceable Main Street—no touristy cack to be found. After a whirl along a river road, I've crossed the bridge over the Yellowstone River and am on the long road that leads to Paradise Valley.

The Chase ranch house is a long, one-story wooden structure built in a curve facing out toward the mountain and range. It is a powerfully comfortable house, one that creaks with thirty years of family dinners and people piling in and out, stomping snow from their boots, dumping loads of wood, canning, cooking, riding, hiking, and spending evenings around the fire reading. I climb a set of unadorned concrete steps to a back porch, thence to a mudroom, and finally the kitchen.

But I don't make it to the kitchen because I fall to my knees in front of dog kennels. Chase and his wife have seven—seven!—Jack Russell terriers just like mine, who are penned so that they don't swarm the visitor. I pretty much have to be dragged to my feet and into the kitchen, where Diana, a slim, beautiful woman, greets me and hands me off to Alston, who guides me into her study because his is being cleaned. The study is wood paneled, and bookshelves line the walls.

Chase is tall and thin, with large eyes and a classic egg-shaped head. He is polite, too, and mild, and after questioning me closely on my own provenance, he divines my concern for the plight of rural dwellers. I tell him that the real country people, working country people I have talked to, particularly those whose families have worked the land for generations, say that the countryside is *not* at risk. It changes and some things need fixing, but on the whole, it's just fine. The threats are external—or externals, as economists would say . . . I trail off.

Chase doesn't exactly explode, but he comes close.

"Of course! Anyone who lives in the country knows it's not dying. Look at the photographic record. One hundred and fifty years ago right here, there was more sagebrush, few trees, less game. And even the best landscapes, in 1870, had been created by Indians burning the forests to create hunting grounds. When the first settlers arrived, research tells us that they could have driven a wagon train across the country without stopping. The land had been cleared that much. All the landscapes of North America were improved by Indian burning.

"I have a much much darker view of the environmental movement now than I ever have—it's been a case of successive epiphanies, the scales successively peeled from my eyes. It is congregationalism, pantheism, the religious elements of Native American environmentalism meeting Hegelian philosophy and Rachel Carson. What we have is an explosive and extremely dangerous mix."

With *Playing God* and *In a Dark Wood*, Chase traced the intellectual history of environmentalism and came up with something so surprising that during the two weeks after I took it on board, his idea—or rather the connections between ideas—changed everything I thought about the natural world.

Chase, like many thinkers, attributes a fundamental schism in Western thought to the battle between Kant and Hegel. Immanuel Kant asserted the dualist nature of man and in *The Critique of Pure Reason* investigated the limitations of reason. In contrast, Hegel, Mr. thesis-antithesis-synthesis, was a monist, an early we-are-the-world type, and by the late nineteenth century, his thinking ruled.

Karl Popper watched the rise of Hitler and the Nazis as a teenager, left before the Anschluss, and traveled with his parents to Britain, where he converted to Protestantism. As he integrated his early experience, Popper began to ask what conditions allowed fascism to arise and quickly came to realize that all the totalitarian philosophies of the twentieth century owe their origin to Hegel.

"In science," says Chase, "the progression goes from Hegel to Ernst Haeckel. Haeckel coined the term *ecology*, from *oikos*, meaning 'living relations,' and *logos*, meaning 'the study of.'" Haeckel was Hitler's go-to propeller head. Individuals, Haeckel argued, following Hegel, do not have a separate existence; they are part of large wholes—the tribe, the nation, the environment.

"The gestation," says Chase, "goes therefore from Hegel to Haeckel to Hitler, or Marx and Engels to Soviet science or Nazi science—which were both a corruption of science. We're really talking about the corruption of science. Hegel allowed the breakdown of the firewall between science and religion. Science, therefore, now came with religious significance.

"And after Lenin's death, the effort was to create a proletariat science. Which declared that *the authorities have the power to change nature*. Genetic change can be forced, we have [Soviet agriculture minister Trofim] Lysenko to thank for that. He brought forward the idea of inheritance of acquired characteristics. The Soviets banned the teaching of Mendelian genetics, put to death the scientists who taught it, and sent the Darwinians to the gulag.

"H. J. Muller—an American fellow traveler—went to Russia to help Stalin breed the perfect proletarian worker. He had a profound influence on Rachel Carson, and his rave review of *Silent Spring* in the *New York Times* sent the book over the top in sales. Everything she wrote was a lie, but she is still treated as a saint."

He's on a roll now.

"While this was going on in the twenties in Soviet Russia, Haeckel's influence exploded in England. The effort to control everything meant molding human nature. The Fabians Beatrice and Sidney Webb at Oxford became infatuated with Haeckel, not as a way of understanding nature, but as a schema for social engineering, for control of societies in their environment. The Fabians were racist; they had an entirely paternalistic attitude to people of color. Oxford—my alma mater, or one of them—invented Hegelian ecology.

"Hegelian ecology is an ecology for the total control of everything," he concludes. "And its tool, the *perfect* tool to effect this control, was the concept of the ecosystem."

In 1935, Oxford botanist Arthur G. Tansley invented the notion of the ecosystem. The ecosystem, he posited, is "the basic unit of nature," more fundamental, he claimed, following Hegel, than the individuals that form its parts. The balance and unity of nature, two ancient notions, had fused and found their modern roosting place.

In 1946, G. Evelyn Hutchinson, a British zoologist at Yale, advanced the theory that the ecosystem was a feedback loop of energy flows that operated to keep the system stable in the face of environmental disturbance. These flows worked like a thermostat to keep the community in balance. A healthy ecosystem is in balance. When affected by disturbance, a *healthy* ecosystem returns to balance. If one part is missing, it cannot be in balance, and therefore it must be unhealthy.

Next, using the concept of a *niche*—the habitat occupied by an organism—Hutchinson decided that biological diversity promotes ecosystem stability. And from the late 1940s on, governments and universities started ladling money onto scientists who worked to explore this set of seductive ideas. In 1959, Hutchinson continued the sophistry, stating that "communities of many diversified organisms are better able to persist than are communities of fewer, less diversified organisms."[1]

It is an idea that we all accept as truth. You do, your mother does, and I did too. Right, left, centrist, apolitical, whatever our stance or lack of it, we all believe that diversity and balance mean ecosystem health, the health of the planet's natural systems. As Chase says, it was an idea that took off like a rocket.

You can imagine, if you weren't there, how thrilling this was to the enthusiasts of the 1960s. Theoretical ecology burgeoned. Drawing on the idea of the "oneness" of nature and the pantheism of Thoreau, Rousseau, Aldo Leopold, John Muir, and Ansel Adams, the ecosystem became God.

But here's the rub: it was theoretical.

By jettisoning Kant and his insistence that science be based on empirical evidence, scientists had explained life on earth using a construction that does not exist. No healthy, self-regulating, and therefore so-called balanced ecosystem has ever been found. Think of the failed biosphere projects, which tried to re-create the conditions for life using the ecosystem idea. Despite the failure to create one, or perhaps because of it, the ecosystem is a lab rat's dream. It sounds right, it has been promoted assiduously, it is widely accepted, and best of all, there is a flood of funding for it. Because they don't exist and therefore no one has found a healthy one, the hunt must continue. It is the perfect racket. As Simon A. Levin, Moffett Professor of Biology at Princeton, admitted in 1985, "What we call . . . an ecosystem is often a fiction, an arbitrary restriction of spatial boundaries, rather than a reflection . . . of species change."[2]

Toward the end of the Clinton administration, environmental consultant Allan K. Fitzsimmons wrote:

> *Sixty years after the ecosystem idea surfaced in the scientific literature; after decades of dominance on university campuses; after thousands of books, articles, conferences, and monographs; scholars cannot agree on the most fundamental matters regarding ecosystems. They do not agree on what constitutes the core characteristics of ecosystems. They cannot say where ecosystems begin or end in space or time, or tell us when one ecosystem replaces another on the landscape. They cannot agree on how to locate ecosystems. They offer no generally accepted definitions or measures of health, integrity, or sustainability.*[3]

"What is sacred in our society today" says Chase, "is the ecosystem—it trumps the individual and individual rights, it trumps God. It's the concept that trumps everything—it's a wonderful tool. If you turn biology into religion, you have created a mechanism for the total control of everything.

"You start with small areas, define an ecosystem, and set it aside. But you find things are changing, bad things are happening; the real enemy is change. Humans are damaging things, kick them out, extend the area. Oh, there's still change; make the ecosystem bigger, till finally you get the Wildlands Project. Oh, that's not big enough. We've already set more land aside than all of New England and New York State. Where's this going to stop?

"We have to find the complete ecosystem, and we haven't been able to find it. So if we can't find a complete ecosystem, it must be the *universe*—the self-regulating mechanism must be Gaia, the atmosphere! You control everything on the surface of the earth. You control carbon dioxide, which is what we breathe [*sic*], and so then we control people's bodies and we're controlling everybody, which is getting closer to what we want. . . ."

We both collapse in laughter.

"But the thing is that in nature," Chase picks up, "everything is always changing. Look at any area, and there is change. There is no such thing as the health of nature—it's in the eye of the beholder, a covert reference to the human body. The *body* is a self-regulating ecosystem. If you take things out, like the liver, the human body dies. But the body is not nature.

"A landscape is a landscape, and you can make decisions on whether to do something with it or not. If you want to manage for deer, you do it one way, and if you want to manage for butterflies and other early successional creatures, you do another. These are deliberate choices.

"You can only speak of healthy for, not healthy. It's not some metaphysical category."

Although the UN's Convention on Biological Diversity was not ratified by the United States, the Clinton administration launched the massive study recommended by the convention, under the direction of the U.S. Geological Survey (USGS). Working from the template established by the Nature Conservancy's Natural Heritage Network, or NatureServe, the study was conducted with the cooperation of all land-management agencies, state and federal, to determine species relations, habitat requirements, and hydrologic functions of every "ecosystem" in the United States. The result of that study is the so-called Gap Analysis Program (GAP).[4]

The GAP process overlays data in an astonishing ten dimensions. It uses vegetation types and vertebrate and butterfly species (and other taxa, such as vascular plants) as indicators of biodiversity. Maps of existing vegetation are prepared from Landsat satellite imagery and other sources and entered into a geographic information system (GIS).[5] Vegetation maps are verified through field spot checks and examination of aerial photographs. Predicted species distributions are based on existing range maps and other distributional data, combined with information on the habitat affinities of each species. Distribution maps for individual species are overlaid to produce maps of species richness, which can be created for any group of species of biological or political interest.[6]

What's the point? The answer is in the name: gap analysis. Biologists and land activists use gap analysis to find ways of increasing ecosystem elements seen as missing and ways of linking existing reserves so as to increase wilderness or biodiversity corridors.

The final step in gap analysis is to superimpose property ownership on those maps.

Land ownership is divided into four classes of stewardship status. Class 1 is fully protected as wilderness areas. Class 2 land is mostly protected, like national parks and many wildlife refuges. Class 3 is partially protected, like national landmarks and multiple-use U.S. Forest Service lands. Class 4, usually private land, has no known land protection, though by 2012, almost all privately held rural land was enmeshed in restrictive environmental regulation.

Classes 1 and 2 are often combined. As illustrated by the astonishing Web video *Taking Liberty*, tremendous political pressure is levied by ENGOs to convert Class 3 multiple-use lands to Class 1 or 2 protected wildlands. Such a change automatically denies timber, range, fishing, farming, recreation, and firewood-gathering resource users from their traditional access to multiple-use federal or state land. In many cases, access to this land is their livelihood. Then any evidence of man's presence on that land is ripped out in order to classify it as wilderness, hence the destruction of relicts of the pre-Revolutionary settlers and dams in the headwaters of every river system.

The amount of land classified as wilderness has grown tenfold since ecosystem theory took flight, growing from 9 million acres in 1964 to 110 million acres today. In addition, the Grizzly Bear Recovery Plan covers 23 million acres in the northern Rockies, and the Northwest Forest Plan protects old-growth forest on behalf of the spotted owl on 24 million acres,[7] bringing us to a total of almost 157 million acres of wilderness or quasi-wilderness area.

Another 75 million acres of wilderness is categorized under wetlands, conservation reserve, habitat conservation, and safe harbor agreements, bringing us to 232 million acres of no-touch and virtually no-access wilderness; various other classifications like wilderness study areas, areas of critical environmental concern, wild and scenic rivers, research areas, and scenic areas bring the formal count, as of 2005, to 337 million acres in designated wilderness. When all states report in, later in 2012, acreage numbers will be updated, and it is expected that many more million acres of protected lands will be included in the count. Instead of the alleged predatory development of rural land, from 1950 to 2000, rural open space in the United States, land lying fallow and bearing no human footprint, has increased from 46 percent to 50 percent of the U.S. landmass. More than one billion acres in the United States remains open space.[8]

In the United States, urban areas take up 2.6 percent of the landmass.[9] Including agricultural and rural development raises the figure to 5.6 percent[10] of the American land base in use. In Canada, urban land, rural development, and farmland take up 3 percent of the landmass.[11]

That leaves a little more than 95 percent of North America undeveloped—hardly the internal map most of us walk around with but nonetheless the truth.

• • •

Splayed in front of me is volume 2 of *The Conservation Manual of the Sensitive Ecosystems Inventory: East Vancouver Island and Gulf Islands*. The 256-page manual is published by the Environmental Conservation Branch of the Pacific and Yukon Region of the Canadian Wildlife Service.

In its pages and those of its fellows, there are strict and furious directions on what each landowner and public land steward is to do on his property to save the ecosystems of which his property is part, and these are duplicated in every single jurisdiction in the world. And if part of the ecosystem is at risk, if regulation is not already in place, it will follow. It is only a matter of time.

The document is by turns romantic and chilling. There are pages of forgotten wildflowers, evoking in the susceptible reader intense nostalgia. And there are forms: Ground Truthing forms for both Uplands and Wetlands and SEI Site Nomination forms, which enable you to nominate a meadow or bog you think needs saving from human use. There are Ecosystem Keys, which show you how to identify an ecosystem, and there's a listing of Organizations and Resources, so that you can find free help to further your knowledge of any ecosystem you wish to protect, and perhaps a little muscle.

Feeling bereft? Wishing life were simpler, easier, more like your grandparents' in a bucolic country town? This is a manual for action on how to freeze "sensitive" land you might find, whether yours or your neighbor's. As Chase puts it, by 1995 "any scruffy activist with a typewriter could stop a timber sale."

Of course, because no one has found an ecosystem in nature, "identified" through keys or otherwise, everything is declared out of whack. Every single ecosystem is at risk. As Chase puts it, "When field studies failed to find projected equilibria, they took this as a sign that ecological catastrophe was imminent."

Chase identifies three wrong turns made by scientists in the grip of a very bad idea. "First, by employing concepts referring to unverifiable holistic entities such as ecosystems and their properties, it was no longer science but philosophy. Second, many practitioners were embracing unsubstantiated hypotheses concerning the balance of nature and the need for biodiversity. And now, third, a growing cadre of environmental philosophers was transforming the doctrine into political theory.

"Not only," Chase concludes in *In a Dark Wood*, "were living things interdependent, they said, but this diversity was also intrinsically sacred."

Fifteen years after Chase wrote this sentence, ecosystem theory is embedded in every environmental policy and every environmental regulation arising from those policies in every country in the world. Especially the poorest: even Ecuador has recently made biocentrism its legislative first principle, against which all endeavors must be measured. Protecting Ecuador's ecosystems trumps the human rights of Ecuadorians. Locking down critically necessary natural resources is more important than the life of a man, woman, or child. Thus a fallacy perpetrated by the most privileged among us can kill an Ecuadorian, having first denied him employment, education for his children, and medical care.

As Chase says and as I am directly experiencing, the first result of the ecosystem idea was a "labyrinthine public planning process, which ensured, as the historian David A. Clary observes, 'the eternal generation of turgid documents to be reviewed and revised forever. Planning became an end in itself.' " Hence, my conservation manual has just been updated, at a cost of hundreds of thousands of dollars, and is replete with ever stricter instructions and requirements, interspersed with barely controlled hysteria over species loss.

The acceptance of a false idea as real tilted the field to do-ityourself activism. Like the Red Queen, anyone who could make nonsense sound like sense by appealing to the kinds of emotions I am feeling in my destroyed but still gorgeous meadow was king of the heap.

As Michael Crichton asserted in *State of Fear*, his takedown of the climate change machine, prior to World War II, eugenics—breeding the superhuman—was the idea that obsessed the *bien pensants*. Adherents included Theodore Roosevelt, Woodrow Wilson, Winston Churchill, Oliver Wendell Holmes, Louis Brandeis, Alexander Graham Bell, Margaret Sanger, Luther Burbank, Leland Stanford (founder of Stanford University), H. G. Wells, George Bernard Shaw, and hundreds of others. Nobel Prize winners gave support. Research was backed by the Carnegie and Rockefeller foundations. The Cold Spring Harbor Laboratory was built to carry out this research, but important work was also done at Harvard, Yale, Princeton, Stanford, and Johns Hopkins. Legislation to address the "crisis" of bad breeding was passed in states from New York to California.

Creating the perfect human by identifying ideal characteristics and breeding for them embraced Lysenko's and Haeckel's theories. The discovery of Hitler's ovens killed that idea once and for all, but Chase makes the point in his fourth book, *We Give Our Hearts to Dogs to Tear: Intimations of Their Immortality*, we are currently and enthusiastically breeding our dogs into tragic genetic misfirings. Breeding for the perfect glossy golden retriever coat seems to go along with a geometrically increased chance of the dog being functionally crippled by the age of seven.

Likewise, attempting to force a perfect balance on an "ecosystem" creates biological disasters on a scale only dreamed of even by the worst, first Rockefeller. If the underlying premise is incorrect, what follows is an error cascade. Until 1995, the earth sciences were proceeding along a course established over the past five hundred years: test, retest, argue, test again, fight bitterly over what is right and what is not right, and then, after a theory has been so thoroughly vetted that it can stand on its own without question, accept it as fact. Adaptive or evidence-based management had worked for five hundred years, and new discoveries poured from the labs of ecologists who worked with higher math, the only calculations that could approach the complexity of nature. Their findings were on small scale. Small was possible. The big, big picture was decades, even hundreds of years out. Adaptive management like German silviculture, the state of the art in forest management, had been practiced in both national and private forests. Yes, there had been mistakes and excessive predation of forest stock, largely by the leveraged-buyout kings of the 1980s, but on balance, the forests of North America were healthy and strong.

Then came Rio. The brainchild of one of the most feverish of the new breed of socialists, Maurice (pronounced like Morris) Strong, the 1992 Earth Summit and Bill Clinton's election managed to run the old-school earth sciences right off a cliff, and not just in North America. Rio must have been quite a party, because it seemed that everyone came back high as a kite, spouting the most ridiculous low-rent intellectual construct ever to be considered by a sane people, viz.:

> *[C]urrent lifestyles and consumption patterns of the affluent middle class—involving high meat intake, use of fossil fuels, appliances, home and work air conditioning, and suburban housing—are not sustainable.*[12]
>
> —MAURICE STRONG, SECRETARY GENERAL OF THE UN CONFERENCE ON ENVIRONMENT AND DEVELOPMENT (AKA THE EARTH SUMMIT), POSITION PAPER #2, 1991

In order to solve this dreadful problem of the affluent middle class, Clinton in 1993 appointed the President's Council on Sustainable Development.[13] As Al Gore had written in *Earth in the Balance*, published the same year, sustainable development was necessary but meant "sacrifice, struggle and a wrenching transformation" of American society. Despite Al's lament that "minor shifts in policy, marginal adjustments in ongoing programs, [and] moderate improvements in laws and regulations" wouldn't work, in fact the only way a wrenching transformation was possible was to introduce the change in increments and out of the view of any gatekeeper in the media.

Who sat on that council? Representatives from the Nature Conservancy, the Sierra Club, the Environmental Defense Fund, the UN affiliate the World Resources Institute, and of course, Dow Chemical and American Electric Power, the two top polluters in the world. Lots of po-faced cabinet secretaries bent on saving the world made up the complement. And Enron was represented there, for its innovative financial schemes, no doubt. Thus all the usual suspects were present, plus a few unlovely surprises. Dow, American Electric, and Enron were certainly prescient to involve themselves in the "visioning" of a massive new regulatory scheme meant to control the exhale of every human on the planet.

To parse the activities of the President's Council on Sustainable Development is no small task. The literature is vast, and as is typical, the words seem to add up to nothing at all momentous, simply a hewing to fairness, care for the future, and inclusion. This was a new value system, it was said. Apparently the founding values of the Americas and Europe, that is, Judeo-Christian values, didn't include fairness, inclusion, and care for the future. Never mind: old values were bad; new values were on the way! The PCSD formed four task forces, one on climate change, another on environmental management (regulation writing), an international task force ("encouraging" sustainable development around the world), and finally a metropolitan and rural task force, which was considered the most successful.

Let's try to slog our way through that "success," which was successful at least in part because it happened far out of the way of any pesky journalist or think tanker with an "agenda." No James O'Keefe was liable to turn up at the Southern Muskoka sustainability visioning thrash to mock the proceedings. No, wherever the sustainability corporals went in the heartland, they were met with proper reverence. Sustainability prose follows:

> *The local, state and regional approach task force appeared to be the most dynamic. This task force undertook studies and recommended new approaches to encourage competing or neighboring jurisdictions to work jointly. Also, it supported several efforts to catalyze local and regional implementation of sustainable development. With the support of the PCSD, four federal agencies (DOC, DOE, EPA, and USDA) provided funding to create the Joint Center for Sustainable Communities (JCSC). The JCSC is designed as an information clearinghouse and technical assistance center that supports innovative multi-jurisdictional approaches to community and urban development and also showcases communities that implement or demonstrate PCSD recommendations. A subset of members initiated a regional council in the Pacific Northwest. Members of this regional council (the Pacific Northwest Council for Sustainable Development) include business, tribal, state and non-governmental organizations from the states of Alaska, Idaho, Oregon and Washington.* This council, much like the national council, developed a vision for the region and began building a coalition to support its implementation.*[14]

* The Pacific Northwest Council for Sustainable Development generated ICBEMP, the Interior Columbia River Basin Ecosystem Management Program, which was so unpopular that it disappeared without notice but not until politicizing tens of thousands of furious landowners on permanent high alert for the next incursion.

And what was that vision, in its essentials? The Global Biodiversity Assessment Report[15] listed the following things as unsustainable: private property, single-family homes, paved roads, ski runs, golf courses, logging, plowing, hunting, dams, fences, paddocks, grazing, fish ponds, fisheries, drain systems, pipelines, pesticides, fertilizer, cemeteries, sewers, and so on.

In 1993, the EPA circulated a detailed action plan on how, over the next eight years, U.S. environmental regulations would conform to those of the UN. "Natural resource and environmental agencies . . . should . . . develop a joint strategy to help the United States fulfill its existing international obligations [for example, Convention on Biological Diversity, Agenda 21] . . . the executive branch should direct federal agencies to evaluate national policies . . . in light of international policies and obligations, and to amend national policies to achieve international objectives."[16]

Could it be any plainer than that? The mission of all the agencies took a sharp turn from assisting citizens in using natural resources to protecting those resources from citizens. It was the claimed success that puzzled, initially. Did people just roll over? What did it mean? What was the upshot, the result, the effect of this amorphous term *sustainability*? Could that success have occurred because bureaucrats in Commerce, Energy, EPA, and Ag already held police powers over everything that exhales and makes money in the country and packed in a briefcase lots of lovely money to give to people with proper respect for their towering intellectual abilities and compassionate agenda? One can easily imagine the full-on enthusiasm for "sustainability" among this particular subset of humans. Or were they still merely human? An internal Bureau of Land Management document of the time declared: "All ecosystem management activities should consider human beings as a biological resource." What fun! BLM employees were clearly more human than others. And the extra-special humans liked it; they got to travel and give seminars where preening was not only possible but required. They got to involve themselves with everything, including the height and footprint of barns, the siting of new houses, the evaluation of chicken houses, and the registering of saddle horses. In fact, as it turned out, the exigencies of creating "sustainability" meant they had to measure the depth and speed of flow in your irrigation system three times a week, put four meters on your wells, and drive down any driveway and be met with a biological resource so servile that one could feel really important for the rest of the day. The overtime on report writing alone could send the whole family to Provence for the summer! Heavens, this was employment on a scale so juicy, so rich, so fantastic, that the more sustainability, the better.

By 1997, the last year of the PCSD, wealth transfer from the "affluent middle class" to the new nomenklatura was fully entrained.

Challenging the foundational myth of the sustainability revolution—the ecosystem—became risky even for the distinguished. And the onerous requirements of old-school science? Mere annoyance.

Citizen Enforcers

A phone call summons me to an Advisory Planning Committee meeting to discuss my application. Heart pounding, I go to the Islands Trust office. New, it is housed in a corner of a steel warehouse owned by the local electric company, retrofitted as a warren of functional rooms. A small lobby decants the visitor toward a counter, behind which middle-aged women glare and snap at a constant stream of petitioners and occasionally smile at one. There is a rock-covered wall, a nod to the mandate of nature protection, a bulletin board of scheduled meetings, and one distinctly inhospitable chair.

In an airless, windowless room, eight members of the APC gather. It is a volunteer committee, the weightiest of many trust volunteer committees; every development application pitches up before them first. I have had the members' names on a list for two months now, debating whether to call them or not, wondering who they are and what kind of animus they might bring to bear on my application. Two well-known activists, Sally Johns and Jackie Booth, dominate the committee. Sally is a forester, and Jackie is a marine biologist. Both are against, in principle, the kind of land-use mechanism I am using to subdivide my property. Three new women members, who may be uncertain of their purpose, are in place, and three men: a Realtor, an architect, and a retired government biologist. The Realtor is the only one who smiles at me. All the others greet me with accusatory glares or spiky suspicion, or they pretend I do not exist.

Brent has joined me, and I take comfort in his solid presence, but I can tell he is nervous too. We sit at the back of the room, away from the table. Cathy, the staff member shepherding my application, is at the table, stolid, unflappable. After the various formalities are out of the way, she presents my application.

A dour silence fills the room.

"This is a density transfer," says Sally Johns, square, healthy, authoritative.

"Yes," says Cathy. Her tone is casual, matter-of-fact.

"I thought we had agreed that density transfers were no longer to be considered."

"Yes, well, we understand that is what you want. But it is impossible to do that without an OCP revision."

"We revise the OCP all the time."

I can hear blood rushing through my head.

"But this revision would have to go before the community," Cathy says. "And the application was in hand before any indication was made by the committee that density transfers were to be turned down."

A rustle of disagreement greets her sentence, but they aren't as sure of themselves as they'd like to be. I silently congratulate myself for sliding in the application when I did.

Jackie joins in. "Does this connect to the Cusheon Lake watershed?"

Cathy, still catching up with the island's geography, pages through her papers for a few seconds, uncertain. I dig my elbow into Brent's waist.

"No," says Brent, obedient. "The applicant's creek empties into Ford Lake, then the ocean. This is Fulford Creek."

Jackie and Sally look at me with new respect. Fulford Creek is a valuable creek, a good creek, a noble creek indeed. And I have promised to restore my part of it. I am a good green citizen. The rushing in my head slows—I have become acutely aware of my blood pressure through this venture. Usually so low my GP says I've got the metabolism of a slug, it has been careening around like an unbroken horse.

"We'll need a site visit," says Sally. The hissing in my head returns.

"How about now?" asks Brent.

They all slowly turn to look at him, shocked. That is out of the question. That would be fast, fast is bad, fast is practically illegal; in fact they are pretty sure it is illegal and turn to Cathy for guidance. She says, in fact, it is illegal: the site visit must be advertised. But Brent's little maneuver means that they do schedule a site visit for that Friday, a week away, enough time to get an ad in the paper. Brent can't make it; can I handle it alone? Yes, I say, at $110 an hour, I most certainly can.

So, seven days later, in the late winter sun, waiting for the people who will decide my fate, I stand, shears in hand, beside a pile of broom six by six by six feet—the only size I am allowed to burn—jumpy and afraid. The cars start to arrive. I bound out, as eager and friendly as a labrador.

The sun's warmth teases, and I have cleared enough broom for them to see the disaster that had been produced by thirty years of neglect—the sod sold, the gravel mined—and the beauty that could return. The committee moves in a gaggle across the stream and up into the forest. As we climb up and up, their breath grows short, as does mine. I point out the year-round spring, something of a miracle in the supposedly dry islands, and herd them over to the lip of the ravine. It is in full early spring spate, a splendid sight.

"Putting my application through means that a covenant will be placed on it, and it will never be touched," I say. Sally Johns makes a gesture of impatience, and I back away like a Tudor courtier.

They stand, stare, then start talking quietly among themselves.

Then finally the sternest among them breaks into a wide grin. "This is gorgeous," says biologist John Sprague—he throws his arms wide. "We will be doing something wonderful that will last for a long time."

I nearly fall to the ground in relief. The quiet women on the committee look at him and nod.

Sally Johns talks about preventing future logging.

"No more logging on this property," I rush in. "It's too messy."

Sally hoots like a large boisterous owl.

"You should tell that to your forestry clients, Sally," says Jackie. "Logging is bad because it's messy." They fall apart on the trail.

This makes me feel as silly as Paris Hilton, but that's fine, I'm an idiot, just say yes, for pity's sake. We climb down into the meadow, talking easily, and then Jackie decides she'd like to visit the other creek. So we clamber a hundred paces up the headwaters of Fulford Creek. It is covered in mosses, and the creek is deep; one waterfall follows another in a series of splashes. They want this creek covenanted too.

I agree immediately, only slightly embarrassed by myself, and despite the fact that they have just spent another $5,000 of my money on a whim—riparian regulations already in place mean that neither of these creeks would ever be touched—I push this out of my mind. Then I start to plead. Pleas are a necessary part of the process. I try to be amusing, but at the same time, humble. I believe in the trust, I confess, trying not to think I have been transported to a reeducation camp. I believe in saving the land; I love this place; I paid the mortgage on it for ten years before I could move here. Jim has just five years of life left; he needs the money from the sale. Et cetera. Desperation hovers around me like a bad stink. But they like it, it is human and personal, and they are sympathetic enough. We descend into the meadow. The whole exercise has become so expensive, I've decided to sell my old house with the spectacular view and build another down on the meadow. I have sworn, not in blood precisely, but made a strong promise to build a green house, an exercise that intrigues me anyway, so I offer to show them the possible house site, and they agree.

And they decide to have an informal meeting on the spot. The house site is warm, the sun is shining, and everyone sits on the ground.

Sally Johns begins. "This," she says, "is a density transfer. We are against density transfers. We want the density transfer policy reviewed; we felt it was meant to cluster people into the villages, rather than have them out here in the rural neighborhoods."

"But . . . ," I say from the ground beside Tom, the Realtor. He is lying on his back, staring at the sky.

"I'm sorry, you are not allowed to speak," snaps Sally.

"So while there seems nothing wrong with this subdivision," she continues, "I think we should once again ask for a moratorium on density transfers."

I throw a panicked glance at Cathy, who steps in, with her quiet voice. "The meeting must be advertised and people must be allowed to attend," she says, her tone final.

Jackie shrugs and Sally mutters, but the others climb to their feet and begin to take their leave, most complimenting me on the place, the forest, and its creeks. I have no idea what their decision might be. The mood seemed generally positive, but Sally and Jackie seemed cross and impatient most of the time, and they clearly run the committee. They would likely comb the bylaws looking for something to torpedo my argument. I held out the hope that John Sprague liked it, but the impulse on the committee defaulted always to not allowing anything to change. He would go along with that. The others—I wasn't sure, but I thought I'd made my case to them. On the phone with Brent, he says, cautious, that it might be good. But that you could never tell.

Chapter Six

The Vigorous Preening of Our Moral Betters

High-desert cowboy country is the most recent front in the war to save us from ourselves. Montana fairly crackles with energy. In fact, the so-called intermountain West is growing fast, newly colonized by that aggressive invasive species, retiring upscale Boomers settling near the national parks. But the invasion also includes middle-class families decanting from the cool cities, looking for a place their values may still be recognized and where schools might educate in the old way. Young activists and utopians migrate here after college, bunk in together, and start agitating for jobs at the Turner Foundation or any of the dozens of other foundations, think tanks, land trusts, and activist groups that operate out here. No surprise—the pickings are legendary.

From the top of Montana through Idaho, Utah, Colorado, and New Mexico, all the way to Texas lies the fount from which all protection flows, which is to say the headwaters of the United States. Controlling the origins of the waters of the American West means you have effective control of more than half the continent. Moreover, what's happening here is set to move south and east, just as soon as the tools are fully tested and ready for export. So wherever you turn in this majestic place, an entirely rejigged America is forming, appearing first as the shimmering mirage of PR "visioning" and hardening into grim reality every year. Animal migration corridors, massive buffer zones for those corridors, energy corridors, conservation easements on private lands, travel management plans restricting access to public lands, tens of thousands of wetland buffers, and smart growth are remaking the country into something quite other.

Still, radio remains a monoculture: Christian—evangelical and Catholic. Mark Levin, "the great one." Dave Ramsey, the money guru (and does he have his hands full): "Eat rice and beans until you pay off that credit card debt!" Rush, of course. In fact, just as Limbaugh finishes his three hours on one station, another station a few miles away starts running his show; he is just that popular. There is one NPR station promoting the concerns of the cultural Left, and this morning, suitably enough, *Tapestry* songstress Carole King is telling us why she is promoting NREPA (pronounced ne-REE-pa), the Northern Rockies Ecosystem Protection Act, and why the act is necessary so that the big charismatic species, the grizzly and the gray wolf, can have their wild corridors. King intones the party line: as the grizzly goes, so goes the human, and the NPR host burbles in agreement.

NREPA, designed to throw another 23 million acres into non-use, the second-largest wilderness designation in history, is loathed and feared by every person who lives in working country. Five minutes of research would have shown NPR's "on-air personality" that 58.5 million acres of roadless areas under the Forest Service's jurisdiction already acts as de facto wolf and bear habitat and that an additional 23 million acres is set aside in the Grizzly Recovery Plan. But apparently asking the hard questions of the movement is not in NPR's mandate. Besides, roadless forest is not wilderness; it is possible that someone could gather firewood from the forest floor on those 58.5 million acres. If he could get access, which is doubtful. But proper formal wilderness is policed like a neofascist state and evidently vastly preferable.

King's Idaho log cabin, showcased on the cover of December's *Architectural Digest*, proclaims her a boutique country dweller. At 7,500 square feet, it is about seven times the size of the cabins, trailers, and bungalows lived in by actual ranchers, and at 128 acres, her spread is about a tenth of the size of a barely productive working ranch.

But King's not a real rancher; she just promotes wilderness and the environment and tells us delightedly about how she and her decorator were soul mates, sounding in this and every other way like a refugee from the Upper East Side. She caps her virtue by showing the *Digest* reporter the indigenous plantings in her garden, gushing about how her hundred-mile view is "spiritual." It turns out that King has listed her ranch, and within a week *Architectural Digest* yanks the story off its website, though it stubbornly remains on the newsstands.

Price?—$19,000,000.* At the same time, King announces her tour with James Taylor. Apparently she needs money.

She won't get it from the land. Prices of real ranches, wherein people ranch livestock for a living, have dropped almost 90 percent in the last four years. Plus, for the superwealthy hoping for a trophy home, her shiny honey-colored-logs aesthetic is twenty years out of date. The Nouveau West is building sleek architectural gems referencing the Old West and its workmanlike materials: local stone, corrugated steel, timber beams. But the corrugated steel has been cleverly blackened, its fastening bolts seemingly hand cast, the beams diamond sanded and torched to dark brown perfection. Floors? Concrete, dragged to lend texture, then polished to look like leather.

In fact, the whole intermountain West has been Hollywoodized. The high mountain desert nestled between the Pacific Northwest and the great lush prairie states has been, in many

* Reduced as of 2012 to $11,900,000.

places, cleared of broken-down towns, rusting farm machinery, and corroded steel sheds. And cattle. And people. In fact, Utah's high desert cowboy country looks like a 1950s Western shot to look as if it's 1880, largely courtesy of Bill Clinton and Bruce Babbitt's 1.7-million-acre land grab in 1996, which locked down 60 billion tons of the most environmentally compatible coal in the world. Subsequent acquisitions, public and private, seemingly piecemeal, have set aside many, many millions more acres.

On the drive through Utah toward Denver, one expects the Lone Ranger to appear at the crest of each hill on I-70 and canter down whirling a lasso and singing "Don't Fence Me In." It isn't until I detour through Aspen—the most expensive place to buy real estate in North America—that I realize that we have the John Kerrys, the Tom Cruises, and their multitudinous gang, with their log palace compounds and their effusive giving to conservation outfits, to thank.

Just outside of Aspen, for instance, lies the Rocky Mountain Institute, the mother ship of all Western utopian enterprises, as reverentially thrilled to bits by itself as one imagines were the first dwellers of the notorious Transcendentalist community, Brook Farm. These are good people who produce rafts of reports and projects and websites—and plans: the Carbon War Room, Project Get Ready, *Winning the Oil Endgame*, GreenFootstep.org. They have worthy project briefings, like *Accelerating Solar Power Adoption: Compounding Cost Savings Across the Value Chain*.

The institute has two addresses. The first is Amory Lovins's house, he being the founder, and his house, elaborately tricked out in solar panels, is used as an example of how energy efficient a building can be. I am as impressed by this as I am by his tax planning. The other address, a long, low ranch house a quarter mile away, sits on the edge of the Windstar Land Conservancy. John Denver's country road has evidently brought him right home here; this thousand-acre ranch was his "gift" to the people of Colorado. One wonders where the ranchers who used to produce wealth on this land for their families and community were warehoused. Their old house, with its soft, modest profile, creaking wood floors, and fieldstone fireplace, is absent the outsize trophies of the New West and more charming for it. The New West is aesthetically perfect and stony cold.

Inside, New West women beaver away promoting Lovins's ideas. All claim the shining credentials of the nonprofit world—conferences attended, papers given, projects completed, degrees and certificates won from various international outfits. In the old chicken shed, a hundred paces from the house, Rocky Mountain's Built Environment team works on energy-retrofitting ideas like designing smart garages. All their projects have clever titles, like "The Smart Garage (V2G)*: Guiding the Next Big Energy Solution." But I have just listened to furious truckers vent about RMI's Transformational Trucking Charrette, which the institute recently shoehorned into a regulatory rewrite. It is impossible to overestimate the shimmering anger, the outright refusal to get aboard the green gravy train, by the working middle class—mostly because they know their lives will be squandered paying for it.

RMI's ideas are druglike in their effect. I remember standing on a street corner in Seattle, years before I woke up in my own green hell, effusing like a fool to Denis Hayes, a founder of Earth Day and director of the Bullitt Foundation, about the Lovinses' and Paul Hawken's book, *Natural Capitalism*. We had agreed at lunch that setting up virtuous loops with regard to resource use made beautiful sense, even if it would cost an additional 15 percent layered on top of every other cost, which sounded just doable in the flush of the Bush years. Today? Not so much.

* V2G standing for Vehicle to Grid.

L. Hunter Lovins, the yin to Amory's yang, was a cofounder of Natural Capitalism Inc., and I had watched entranced at West Coast Green in San Francisco as she put on an astounding show demonstrating the glories of green building, explaining that one-third of energy use, two-thirds of electricity use, and between 40 and 90 percent of carbon emissions originate in our shelters—most of it, apparently, completely unnecessary. She turned me into a true believer within minutes with a breathtaking tour through the buildings she and Amory have transformed. She extends her reach by educating senior decision makers in business, government, and civil society "to restore and enhance the natural and human environment." Lovins, in other words, speaks to those with apparently limitless public money to green what the Lovinses think needs greening next.

As they say in the enviro mapping movement, "ground truthing" to come.

Amory Lovins, who recently transformed energy use in the Empire State Building, has announced his retirement from the institute because he is putting all his time into his "most ambitious challenge" yet. "Reinventing Fire" will drive the transition from fossil fuels to efficiency and renewables. He thinks we can get off oil within forty years, but that the project needs his full attention. How lucky are we?

Montana and Colorado are so very lucky. Ted Turner[1] is their largest private landowner, the largest in the whole West; in fact, the largest individual landowner in all of America, until John Malone came along and trumped him by 200,000 acres. Turner owns 2 million acres of ranch land and, according to his boostering staff, has built the biggest body of environmental work anyone has ever seen. Martha Stewart visited with her camera crew late in 2009, and so we can ride his range, talk to his stewards, and watch his vigorous all-absorbing preening. Ted Turner has a lot of ranches—fifteen—and a lot of staff, says Martha, and every single member loves the Ted.

Join us on their ride.

[Guff about Martha's beaver cowboy hat custom-made for her.
Guff about ranch's name (Snowcrest), acreage (12,000), and recycled wood used in house building, guff that the ranches will never be developed and will always be pristine.
Guff about saddles and quarter horses.
Guff about Ted's largest ranch being 590,000 acres and that it crosses the border of Colorado and New Mexico.
They set off.]

MS: This land acquisition project, when did you begin?
TT: In the 1970s. I bought a plantation in South Carolina and my first three bison; I liked them so much, I wanted to see if I could bring them back, so I bought the ranches so they could have a home. We went from three bison in 1977 to fifty thousand today.
MS: You are the largest landowner in the States.
TT: Har har har. Only in acreage, not in value.
MS: But in open space and beautiful open space, you have done a lot to preserve America's most beautiful terrain.
TT: Yes! Yes! No question about it!
MS: It must make you feel real good.
TT: Yes! Yes it does!

[They cross the Ruby River.]

MS: The Ruby River! My first time in the Ruby River! [*Sighs of pleasure*] Good boy, good boy, good boy [to the horse]. Very nice. You sink right into the mud. Very soft.

TT: It's not much, but we call it home.

MS: It's gorgeous. [*Sighs of pleasure*] What animals live on the Snowcrest Ranch?

TT: Mule and white-tailed deer, moose, antelope. Those are the Snowcrest Mountains. It's really pretty when there's snow on those mountains in the spring.

MS: I must compliment you. I have never seen a land so pristine!

TT: When we bought it, it was covered with so much trash—rolls of barbed wire, bottles, and cans. The people who had it before us didn't take care of it. We have been cleaning it up. [*Sigh of pleasure*] It's hard to imagine it being any prettier.

And so endlessly on.

Then, happily for our purposes, Martha sits down with the head of the Turner Endangered Species Fund, Mike Phillips. Mike is an attractive guy—white hair, outdoorsy look, but with suspiciously smooth, rather than weathered, skin. Martha describes Mike as working with the Turner family to "counteract what's going on everywhere, that's going on elsewhere in North America and in the world, really."

(It is useful to know, going in, that if you live outside the megacities and you ever attempt to do anything on or in nature, or indeed anywhere not paved, you understand why the Endangered Species Act is considered the most powerfully restrictive law in the land.)

Mike Phillips: Ted began the Turner Endangered Species Fund in 1997. We thought we could save nature by saving endangered species. Ted has done more than any other man for endangered species, more than anybody on behalf of the environment. It's mind-boggling to be a part of Team Turner.

MS: This is one of the most pristine environments I have ever seen, and as a result, the animals seem very happy, soaring eagles, hawks, deer, happy to be living in a place free of human problems.
MP: I'm not surprised at what you saw. But the species you saw are pretty easy. Deer are pretty easy to live with.[2] Where Ted's interests take him is also welcoming the species which are difficult.

MS: Like the gray wolf.
MP: Exactly. You know, historically, the gray wolf was the most common mammal in North America—they were everywhere—prairies, swamps, mountains, and we took this very common mammal and over two hundred years drove it to the brink of extinction. And that's what provided the justification for reintroducing the species to Yellowstone National Park. And since they've done well[3] in the park, subsequently they've been expanded to places like the Flying D Ranch and Snowcrest Ranch. It was all about righting a wrong. Ted said, We will fit the Flying D, we will fit the Snowcrest Ranch. We will not make the ranch fit us.

MS: That's is such an admirable . . . do you think that this is a question of morality?
MP: For Ted it is.
MS: Because for so many farmers, if they lose a calf or they lose a baby sheep, they're really miserable, and they hate the predator.*

* Just one example here: The Deadman Ranch had historically produced 85 percent calves, and the Haught family could survive on 75 percent calves. The San Francisco pack, after decimating neighboring herds, moved onto their ranch, and Fish and Wildlife rereleased the Ring pack. Their calving has dropped to a 30–50 percent calf crop—either the calves are eaten or the mother cows are too agitated to conceive. Within two years, the Haughts' dream was dead and their ranch was for sale. Outfits like Defenders of Wildlife offer compensation, but, in fact, compensation is paid somewhat sporadically, and that's the polite way of saying it.

MP: Ted's holistic vision, which he has had since he was a boy, has driven his work in a big way. A perfect example is our project on behalf of the Bolson tortoise[4]—it's a big animal, at a hundred pounds; it's a close relation to the Galapagos tortoise. That species has not existed in North America for thousands of years; in fact the only place it exists is in a small spot in central Mexico. So we decided to focus on this species nobody else was, and we are now involved in a substantial project on behalf of the Bolson tortoise that aims to use reintroductions to bring the tortoise back to its prehistoric range and population.

MS: [*Effusive praise*]

MP: We have had black-footed ferrets in captivity for some time, and we've released them into the wild. We have a captive breeding project for Mexican wolves in the Vermejo Park Ranch, New Mexico. Unfortunately some of these species are so very rare that captive breeding is part of what we do.

MS: [*More effusions*]

MP: Most of the credit goes to Ted. He's out there, speaking to people, spreading the message that we need to leave it a bit better tomorrow than we see it today.

Isn't the Ted the greatest? By the way, only a sociopath does not want "to leave this world a little better tomorrow than we see it today."

Oh, remember the ferrets. And the tortoise. And the "immorality" of ranchers shooting wolves.

Robert Redford—with whom I fell in love at ten—is another prime mover of this radical makeover of cowboy country. He raised his family in Utah, and by the 1980s, he apparently got some kind of religion and started to drag his vast fan base into his way of seeing the world. In the 1990s, he was vigorously retailing the New West, first with the Sundance Festival and then through the Sundance Channel and the Sundance gear catalog and retail outlets. Its aesthetic is the "modern take on rustic style" with reverential bows to fishing gear, the hunt, the chase, the kill, leather, wood, fur, rock, blanket, and rusticated tableware. Followed to its logical conclusion, all outdoor activities would be imaginary rather than real but available in a cinematographic way to those with a few bucks to spend in the Sundance conglomerate. The large, boisterous families of ranching country that actually require and in fact invented all this clobber? Not really wanted. Actual rural life is irresponsible, must be eradicated, and Redford and his crowd are funding the charge.

Redford blogs on the Sundance website, touting the fantastic energy potential of the New West, the "hundreds, if not thousands of leases for wind, solar and geothermal energy" that have been taken out on the newly liberated lands. He brags that his "friends" at the Natural Resources Defense Council have mapped it all, too, and identified the precious habitats and the likely candidates for the fantastic cornucopia of energy gold that will gush from low-impact, nontoxic energy projects.

It's a cornucopia for the great and the good, as is almost entirely the way with all things green. As of September 2011, the Ted had received $4.7 billion in loan guarantees for his First Solar investment project, and the owners and managers of Google shared in a staggering $2.142 billion in loan guarantees for their various green energy investment projects. T. Boone Pickens demanded a subsidy of $1 billion a year for five years for his massive wind project before it collapsed into the desert. And Robert F. Kennedy Jr. along with the owners of Google received a $1.4 billion loan guarantee from the Energy Department for their struggling BrightSource project. When that amount was revealed in

November 2011, Kennedy, who seemingly has only one speed—spittle-flying fury—decanted his rage in a *Huffington Post* blog piece titled "Big Carbon's Sock Puppets Declare War on America and the Planet." According to Kennedy, oil money is at the root of all complaints about America's dysfunctional green energy industry, and many green energy companies show promise. Certainly, one devoutly hopes, they will thrive, someday. But the uncomfortable facts remained, and were reported as fact, across the political spectrum, from Breitbart's Big Government to the *Daily Mail*, *Newsweek*, the *Daily Beast*, and the *Sacramento Bee*. Eighty percent of green energy loans had been awarded to Obama funders and associates,[5] and many of those companies were either struggling or heading toward bankruptcy. Federal loan guarantees had allowed some investors to take their companies public, creating new green fortunes long before green energy had been created. New economic paradigms are enormously difficult and expensive to create; lucrative, in general, only for the creators.[6]

"I have faith in the American people, if you get to them and tell them the right story," said Redford in 2009. Mr. Redford has mastered the ethically compromised financial arrangements of environmental activists, campaigning against Big Oil and pipelines while appearing in ads for United Airlines in 2008 designed to promote the romance of international air travel. The "right story" always manages to benefit this bunch. As reported by Ann McIlhenny and Phelim McAleer, Redford is one of the main opponents of a plan by Pacific Union College to build an eco-village in Angwin, California. "The college says it needs the funds because of a dire financial situation. The village is close to Redford's vineyard in the Napa Valley. However, whilst publicly opposing this development 'to preserve the rural heritage,' Redford has quietly been selling development lots in the Sundance Preserve for $2 million. These lots are intended for vacation homes close to Redford's Sundance ski resort."[7] Increased car traffic was the sticking point for the eco-village, which was canceled. The occasional private plane landing in the Sundance Preserve? Not so bad.

The crazy profits lavished on Redford and his colleagues in the film business and biogeoclimatic hedge fund world may have skewed their notion of the way things work in the world of actual dollars per actual megawatt-hour, which is the way we generally wholesale-price energy.

The U.S. Energy Information Administration estimates the subsidy cost of wind runs at $23.50 and solar burns through a bit more at $26.00. Natural gas and petroleum, 25 cents. Coal, 44 cents. "Clean" coal, refined to environmentalists' standards, $29.81. Little wonder Redford can "report" on clean-energy applications being made for the use of the vast public lands of the intermountain American West. Tens of thousands of glittering green jobs to feed the displaced ranching, logging, and farming families were sure to follow, they thought. Reality is a tough mistress, though, and while all the number crunching hasn't been done yet, as of the end of 2011, if the loan guarantees are called upon and a substantial number of the companies have either already declared bankruptcy or cut staff, Mr. Obama's green jobs will have cost the public purse more than $5 million each.

So it was that, after my thousand-mile drive through cowboy fantasy land, I needed numbers that actually added up. I kept driving, through Colorado up through Wyoming to Bozeman, a town of about 25,000 in Montana, host to one of the several branches of the University of Montana. PERC, the Property and Environment Research Center, sits on Analysis Road, just off Research Drive. A number of future-forward institutions are housed off Research Drive, in fact, including the Turner Foundation, spread out in an enormously tasteful New West structure, copper cladding swiftly moving to patina, dark fir board and batten, barn doors with blackened steel hardware. The Turner building has the profile of a sheriff's office, circa 1890. A big, big sheriff's office.

PERC's is not as fancy, but it's pretty nice for a think tank, the decors of which usually, in my experience, ignore aesthetics. Nevertheless, PERC is a real think tank in that it does scholarly work focused principally on data mining. Dull as dirt, but a relief.

Its purpose is to advance theories of how the environment is best used to its and our health and wealth. The fellows at PERC generally find that property rights advance health and wealth better than collective or public oversight. *Case Studies in Forest Management*, *Creating Marine Assets*, *Averting Water Disputes*, *Conserving Biodiversity Through Markets*, and *Conservation Easements: A Closer Look at Federal Tax Policy* are not heady reads. But they do leave the reader excited because, blessedly, knotty problems are laid out, sorted through, and analyzed; without recommending a vast remake of any sector of the economy, the authors then proceed to vigorous problem solving.

PERC, in other words, stays out of the utopia business. If pressed, you could say that they are property-rights utopians. The institute follows the work of Julian Simon, the free-market economist from the University of Chicago who argued that, rather than growing ever more scarce and expensive, resources become less costly as a culture becomes richer. Given the rule of law, free-market conditions, and property rights, said Simon, the environment will be fine. PERC is the natural antagonist to the environmental Left, RMI, Hollywood, NPR, et cetera, though for the sane, PERC is a valuable resource with a strong hold on the meaning and value of conservation.

I like PERC because its founder and director, Terry Anderson, had, in one of his essay collections, introduced me to the environmental Kuznets curve. Much loathed by the environmental Left, the environmental Kuznets curve, first arrived at in the early 1990s, represents the theory that as a country grows richer, its environmental quality first declines, then improves. At $5,000 average annual income, for the sake of argument, water quality shoots up.

In fact, the environmental Kuznets curve proves true. Since the 1970s, water and air quality have improved radically in developed countries.[8] Likewise, controls on noxious chemicals used in every sector are increasingly enforced every passing year. While there may be more to do, there can be no argument on this: by the measurements by every legitimate body, every economic lurch forward is followed by a cleaner environment. However, in countries pulling themselves out of subsistence farming, environmental quality sinks for a few decades. Of course, the strong-arm tactics of the United Nations, the International Union for Conservation of Nature, and the international conservancies prevent those countries from pursuing industrialization because of the environment toxification that accompanies initial development and growth. In so doing, the most privileged people the world has ever seen mire the least advantaged in eternal poverty.

According to Kuznets's critics on Wikipedia, the environmental Kuznets curve does not hold when you consider wilderness, land, and biodiversity protection; it's more N-shaped. As people grow richer, they eat up more land and destroy more species, and biodiversity collapses. This belief lies at the heart of the metastatic growth of conservation in wealthy countries, from people on both Left and Right. But evidence demonstrates conclusively that as soon as people can afford to conserve land, they choose to do so instead of continuing to pave paradise with more strip malls and subdivisions. It is ideology that has turned that genuinely righteous desire into a curse. George W. Bush, for all his perceived faults, was an enthusiastic booster of conservation; conservation of land by the private sector doubled under his watch.

I want to know from PERC, is this right? Is the displacement or impoverishment of rural people necessary? And does that displacement mean that the environment is improving? And I'd like hard evidence.

Bozeman butts up against Yellowstone, so there are hundreds of hotels, some very decent restaurants, jewelers, bookstores, haberdasheries, and vintage clothes shops rivaled only by those in Aspen. I had appeared at the institute expecting only to rifle through its library and talk to a junior fellow or two, but instead was shuffled out the door by Linda, PERC's communications director, with the happy news that Terry Anderson said he'd like to see me, but tomorrow, because he was hunting today. That's why I became aware of the establishments on Bozeman's shopping street.

In the meantime, a freak storm has dumped two feet of snow on the town, and this morning it is snowing so hard that my GPS can't map my position. Snow slides off the roof onto my windshield, forcing me out of the car into blowing ice and snow to clear it. I slide wildly around the deserted campus looking for directions. Finally I arrive, miraculously not late, but in a flop sweat, which increases steadily over the course of a distressing conversation.

It is clear that this is Terry Anderson's fiefdom. His office is less than 250 square feet, with twelve-foot ceilings, necessary because about two dozen dead critters large and small hang from those walls. A massive elk head with twenty-four-point antlers is supported by a steel rod through its neck. There is a warthog—equally mythic, its bottom teeth and jaw a terrifying sight. I cannot take my eyes off it, while Terry explains that warthogs panic so easily it is nearly impossible to shoot one. Therefore the warthog is an early warning system for other animals at the watering hole—something about the tail going straight up when predators are around. I nod distractedly, for I am on to a different panic. Apparently my netbook has frozen. I can neither record him nor take notes, so I natter on, praying that the computer will warm up before he insists on starting the interview. Mercifully, it does.

Anderson has the sleek, well-fed, but extremely fit look of a man who spends time in the bush but comes home (somewhat like Thoreau to his mother's suppers) for a good dinner and a soft bed. We spar a little. I amuse myself by thinking that his blood is up from yesterday's kill. I explain my subdivision on Salt Spring and how preposterous it is, the other absurdist fixes experienced by other islanders, the pitiable state of those who actually live and work in the country, and he shrugs.

"No one cares," he says.

"But . . ."

"Our values have changed. We don't care about rural people; things have moved on."

"But . . . !"

"Look, last week our county commissioners decided to make the highway into the town a billboard-free zone. Which means that property owners no longer have the right to make money by renting billboard space. The property owner with a billboard has lost something, because society as a whole says we don't want it. That is a redistribution of property rights, no question about it. The question we ask at PERC is, Did we get something of value that's worth more than the cost we imposed? The county commission told the property owner he had three more years, decided that was unfair, and extended it to ten. I say tax the people and buy the billboard, unless, of course, it's worth a million dollars a year. In that case, it's probably too expensive."

"But you and I live in boutique country," I point out. "We have the extra money to buy billboards. In working country, real country, we impose onerous costs on rural people and don't pay them."

"Yes," says Terry. "Does this impose costs on rural communities? No doubt it does, but what do we get in return?"

"Isn't this systematic abuse?"

"Yes, because rural economies are generally in decline and not well organized, so they can't argue against an extremely well organized and powerful minority like the environmental movement. They're too diffuse; they can't fight this stuff very well."

"If this powerful minority knew what they were imposing on rural people, would they be quite so quick to impose the costs?" I ask.

"Yes, because they say, You bear the costs, I bear the benefits. I don't care if you can't have pigs, I don't have pig stench. I don't care if you lose your billboards and Elizabeth can't tell where Safeway is when she comes to town. I know where Safeway is; I don't want to look at billboards."

"Still not working country," I argue back. "What about Lander, Wyoming? They've just been told by the BLM and Fish and Wildlife that they have 85 percent of the threatened sage grouse habitat territory and must restrict farming, ranching, mining, drilling, and any development. That decimates their tax base and impoverishes their population. The feds are essentially telling them to take one for the team, because if they don't, the Wilderness Society will sue to have the species federally listed and all of Wyoming will go down. Not only that, Fish and Game, in a conciliatory gesture, implies the shutdown won't be for long, by saying, 'The more we study the grouse, the more we realize they're everywhere. So sign off, and maybe we'll lift it later on.' "

"We don't care. We want to think of Wyoming as pristine. We don't like the poisoned water ponds left by coal bed methane mining. We don't like the destroyed vegetation, the dying wildlife, the poisoned aquifer. We don't care that we need the coal."

He shoots a look at me to see whether I am shocked by the image of pollution of coal bed methane. I am, but I work hard to keep my face neutral. He is amused.

"Look, I don't doubt that the regulatory processes that we put in place to produce the environmental goods that we want have taken a toll on the economy generally and the rural economy in particular. Telling the story that rural communities are being harmed may tug at the heartstrings of rural people, but no one else will care."

Linda Platt, his communications director, jumps in. "You see, what the sage grouse is about is, they want to stop drilling in beautiful Wyoming. That's the hidden agenda. These people are from Chicago, L.A., Atlanta. They won't even drive through Wyoming, but they want to think of Wyoming as beautiful."

"Take the spotted owl case," says Terry. "One of the people instrumental in shutting down the forests told me that 'if the spotted owl hadn't existed, we would have had to invent it.' The goal was to stop logging. And yeah, we stopped logging, in a pretty sketchy way. It is totally questionable whether owls were endangered by logging. Was it good for the overall health of forest? Probably not. Was it good for the spotted owl? It probably didn't make a difference. Did it hurt the overall economies of the West? Yes. Are the people who pushed it going to repent? No way! They think they did God's work, by golly. They stopped the logging."

But is it a higher good or a bloody mess?

One of PERC's fellows, Holly Fretwell, had just published a small, explosive book called *Who Is Minding the Federal Estate? Political Management of America's Public Lands*, a book I read over two nights in Bozeman's antique Holiday Inn, my eyes widening, shocked to the point of leaping up, pacing around the room, wishing I still smoked. This to me was the most important analysis of the effects of environmental activism on rural America to date. But this morning, it appears that Anderson doesn't seem to have quite taken Fretwell's work on board. Otherwise, he wouldn't be quite so insouciant, so playful, while I am ready to march on the Forest Service *and* the Turner Foundation. Fretwell's work made it clear that everything, *everything*, we have been doing was wrong.

Fretwell data-mined in the public research libraries of the Forest Service, the National Park Service, and the Bureau of Land Management to discover the effects of Bill Clinton's sustainability revolution and discovered, from number-rich *audits*, just how those fifteen years of determined sequestration of public lands had affected the lands. Fretwell wasn't interested in the human population at all. She was concerned with the health of the trees, the earth, waters, wild animal and plant life, the range, and grasslands. How had they fared? What had happened to the forests that the Sierra Club was so proud of shutting down?

It was disastrous. Because thinning, salvage harvesting, cleaning up deadfall, et cetera, are expressly forbidden by environmentalists, between 90 and 200 million acres of Western forest is considered *by the Forest Service itself* to be in immediate danger of exploding in a once-in-a-millennium fire that would burn so hot that not only would the seeds in the soil die, but also the dirt itself would be burned to dust.[9]

Charred mountain slopes and stream banks devoid of vegetation lead to increased spring runoff and lower summer water volumes. No single forest practice—neither timber harvesting nor road building—can compare with the damage that wildfires inflict on fish and fish habitat. No single forest practice destroys endangered species like runaway wildfire. The threatened loss will kill more wildlife than industrialization has managed to extirpate in the past hundred years.

Because there is almost no logging permitted on public lands, the Forest Service loses almost eighty cents on each dollar it receives in subsidy. Federal forests have $5 billion in maintenance backlogs. The Park Service has so much land, it can't take care of it and loses eighty-eight cents on every dollar. *In Yellowstone National Park, sewage bled into native trout streams*, and regulation had grown so complex that Congress had to pass a law before the problem could be addressed.

The spotted owl was dying anyway. First of all, its prey was being eaten by the larger barred owl, which had been moving west for the last hundred years, but its supposed natural habitat was dying. In eastern Oregon and Washington's Blue Mountain forests, 6 million acres are dead and dying. The Shasta-Trinity National Forest—formally designated spotted owl habitat—has so much root rot that it is called the Valley of Death. One breeding pair remains.

Speedy removal of trees could have stopped the pine beetle and budworm, but environmental assessments took so long, the pine beetle spread until it was unstoppable.

The now overgrown forests reduce water flow to communities, farms, and ranches below the forest by as much as 50 percent. Deadfall piles reach more than twenty feet high, creating barriers that force large predators down into human communities to look for food.

The elk and antelope are gone from forests in central Idaho. The dense forests, without meadows created by natural fires or logging, are in late succession and beginning to die. Despite the propaganda, nothing thrives in an old-growth forest. Even the owl breeds in clear-cuts. Furthermore, other bird populations that depend on early successional forests are declining.

The Pacific Northwest forests are four times as productive as forests in other parts of North America. By shutting down the most productive forest in the United States instead of logging those 20 million acres, 80 million acres will be logged in less productive forests.

The Bureau of Land Management, which manages a whopping 264 million acres of the American heartland, is so ill managed that for every dollar earned in grazing fees, it loses almost eight, and therefore has no money for maintenance. As the movement has forced cattle off the range over the past twenty years—Cattle-free by '93, Nary a moo by '92, as the bumper stickers read—range fees, which are calculated in animal unit months (AUMs, the amount of forage needed for one cow or calf), dropped precipitously. Fretwell found state rangelands better managed than federal rangelands but did not dig into the health of preserved or set-aside rangelands. Ranching, she mentions almost as an aside, has largely collapsed as a secure long-term way to make money; those ranches left were often forced to diversify into hunting and fishing preserves; and Fretwell mentioned the action of land trusts buying development rights or conservation easements from ranchers as a positive.

Her most important finding to me was the cost of maintaining preserved lands. From 1965 to 2002, the Land and Water Conservation Fund (LWCF), a leading source of funds for land acquisition, provided nearly $12.5 billion for acquisition. But the costs of managing the land were $224 billion, or seventeen times the cost of acquisition. About $10.3 billion in maintenance was spent by LWCF in 2002 alone. If we take the acreage preserved in the continental United States alone over the past ten years as hovering somewhere around 650 million acres, the amount of money to maintain those acres in a state of health simply boggles the mind. Mothballing those roughly 80 million acres of Western forest on behalf of the spotted owl from 1993 to 1996 meant no maintenance was allowed at all. The Forest Service didn't have the money, and any mitigation they proposed immediately landed them in court, fighting the movement's lawyers. Little wonder those forests were dying. I tucked away the question of preserved rangeland health for later.

To Anderson, I praise Fretwell's work, listing some of her findings and admiring the contribution of Anderson's think tank to such careful research.

"Yes!" says Anderson, with no little satisfaction, glad, I imagine, to be back on the side of the angels. "*That's* the kind of work we do at PERC."

As I slide my way across his icy parking lot, still unnerved by Anderson's frank assigning of rural workers to the slag heap of history, I wonder to myself, what if country people whose family and economic lives have been destroyed by our "higher" values are themselves the natural, and indeed the only truly effective, stewards of the lands upon which they live?

And You Thought the Law Was There to Serve You

Brent calls me on the morning of the town meeting. "There is a problem," he says. "There will be opposition from the Neighborhood Association and the Water Preservation Society."

At the meeting, my oldest friend on the island, Susan Russell, sits beside me and knits, and Peter Vincent, the gym owner, flanks my other side. During the meeting, I dig my hand into his pocket or clutch his forearm. He doesn't flinch, his expression serious, even borderline threatening, which I like—especially when I note the nervous glances thrown my way from David Essig and chief planner John Gauld, both of whom seem weedy by comparison.

Peter Lamb, the head of the Salt Spring Conservancy, passes me a letter from the conservancy requesting that my application go first to the Environmental Advisory Committee for assessment and comment before reaching consideration for the first draft bylaw. This is a feint so the EAC can make the point that density transfers must stop, beginning with mine, and slow down the application for a couple of months. He stands up to read the letter. David Essig thanks him.

The Water Preservation Society sits in a long, frowning line. The meeting moves like gelid ice. Agenda items now have times attached to them, so that applicants or objectors can make their efficient arrivals. Just before discussion of my application is to start, fifty people stream noisily into the room. I want to dive under the chair. But Brent Taylor arrives too and sits beside Peter, at the end of the front row, close to the head table. Cathy introduces the application. Hands shoot up.

Teresa stands up first. She is tiny, fey, with clouds of dark hair surrounding a clenched little face. She is a recent transplant from Los Angeles, where she was an environmental supervisor with the Los Angeles Environmental Affairs Department. She is at every meeting complaining about one thing or another, suggesting restrictions and more enforcement at every turn. Her arms are braced with black athletic bandages and her ankles wrapped as if she were an athlete with an injury. She limps forward.

"The Water Preservation Society is very concerned about this application. We think it is going to impact the Cusheon Lake watershed."

Brent parries this by saying no, my property is two watersheds over.

Heedless, Teresa charges on. "We are also concerned that the house is going to be sited too close to the creek."

"The house," Brent says, "will be sited as required by the development permit regulations and the best management practices of the Water, Land and Air Ministry."

Twenty hands shoot up. The trust officers visibly shrink and sigh. One woman gets up to say, "We thought density transfers had been stopped, and we wonder what the community amenity is in this subdivision."

Kimberly Lineger, one of the local trust members, explodes. "The community benefit is in the density transfer. The island made a moral and financial commitment, when the province paid for part of Burgoyne Bay Park, to buy those densities and pay the government back. There are seventy-three densities, and after this one goes through, there will be seventy-two. If anyone wants to put up the money, which would be about two million dollars, we would be happy to retire them."

Crashing silence from the Water Preservation Society. I admire Kimberly's deft hand.

At the back, yet more hands shoot up. David Essig sighs heavily and looks at the audience. "There will be a public hearing on this, and you will all have a chance to make your objections and ask your questions then. Who will move that this bylaw draft be passed?"

Eric Booth moves, and Kimberly seconds. All three officers vote to pass my application into draft bylaw, and I am through to the next stage.

Three more bylaw readings are required to finalize the rezoning, and none of these is a lock. At the end of the week, the *Gulf Islands Driftwood* reports that the Water Preservation Society and the Neighborhood Association plan to vigorously oppose my application.

CHAPTER SEVEN

It's Not About the Salmon

In the fall of 2009, the *New York Times Book Review* gave its entire prized front page to *Methland: The Death and Life of an American Small Town*, an adrenaline-rich journey into desperation on a gruesome scale—one subject, during a full-blown house fire, wouldn't leave his crank, despite seeing the flesh melting off his arms. It was an unstoppable read, and it went through seven printings in hardcover alone. Nick Reding took four years to drill deep into Oelwein, Iowa, recording the town's spiral into violent crime, family breakdown, the rolling up of businesses, then its gutting. Much of the disaster was triggered by meth use. In many rural areas, meth cooking is the only growth industry. And the meth market attracts Mexican gangs, who are especially violent and who have set up anywhere vulnerable.

Winter's Bone, one of 2010's nominated Best Pictures, demonstrates to what depths some rural lives—once possible, even healthy—have been reduced. A seventeen-year-old girl, the only support of her siblings and catatonic mother, tries to find her cooker father (or his body) before the bondsmen take their house. Tracked out and multiplied many tens of thousands of times, *Winter's Bone* demonstrates the powder keg lying behind the freeways and strip malls of the heartland. Dismiss it as "something happening over there where it doesn't matter" and you miss an important dimension of modern life.

Nick Reding identified the underlying reasons for the grand-scale incursion of meth into rural life as the collapse of American manufacturing and the consolidation of every niche in the food production chain. The four or five operators left squeeze producer prices mercilessly. The result is the vanishing of opportunity and independence in small-town America. It is so true as to be axiomatic, a trope, a wheeze—farms and ranches are consolidating under corporate control, and people are streaming into cities worldwide. Fact—and maybe a good thing.

Except for those Colorado Meth Project billboards that come as a visceral shock to anyone driving I-25. Nobody expects to lose their virginity here, punk graphic language splashed over a grainy black-and-white photo of a filthy toilet. Meth will change that. Okay, for the hyperrationalist, perhaps a good thing with collateral damage. Regrettable, as Terry Anderson points out, but no one cares.

No one expects to spend a romantic evening here. A black-and-white photo of a prison cell with filthy ticking mattresses on rusting bunk beds. Meth will change that.

At least we have to ask if it is necessary.

Winter's Bone makes clear that most rural people, spared the restlessness of youth or the ambition that comes with atypical talent, love their families and do not want to leave. They prefer their lives, even desperate, to those lived in the big cities. Nature isn't escape to them; it is the singing verdant juice of their existence. And those families, as degraded as some have become, are tight, loving, and profoundly meaningful, certainly as meaningful as those living in condos with a loaded cable box in an L.A. or Kansas City suburb.

"These people make me so sick, I can't stand it."

"In Minnesota, where I am, they come in here and say, 'The people in these rural areas, they need help.' No, we don't. Leave us alone."

"I tell people that they should have known that when they started killing off the horses because there's no water—there's one horse for every twenty-three hundred cattle by the way—you were next."

"The BLM [Bureau of Land Management] goes onto a rancher's land to count the heads of cattle, then sends outrageous bills for cattle counting, and because the rancher can't pay the BLM, they take the cattle. That's happened to several ranchers in Colorado. They're cleaning the West out, that's what they're doing."

"The country is being liquidated."

"In Idaho, a federal permittee was told he would have to bypass water to protect two aquatic species and find another source of water. The BLM follows the multiple-ruse doctrine; they just make up any reason."

"Farmers with wells for household use, for irrigation, they are coming out and telling them we have to put a meter on it, because the water doesn't belong to you, it comes from somewhere else."

"You do not take my waters. That's a line in the sand."

"Cutting off someone's water is a physical assault."

"If you don't have water, you are dead. Your livestock is dead, your land is dead, you can't grow food, your crops are dead; it's genocide."

I am sitting in on the *Truth Squad*, a blog radio show out of Farm Wars and the PPJ Gazette.[1] The two women who run the show, Marti Oakley and Barb Peterson, are the kind of women you run to in trouble—real trouble, like earthquake or floods—big-boned, big-muscled, and strong-minded. After talking to them, I think the adage that America is a sleeping giant bypasses the men. When you wake up the women, you'd better watch out. These are not the mewling rights junkies of the cities; they are furiously independent women who can finagle a steer into a truck, shoot a wolf tearing up their lambs without a hysterical phone call to Fish and Game, breech-birth a calf, and feed large families on $150 a month. And they are fully woken up, informed, mad as hornets, and just getting started. Today the guests on the Truth Squad are two young country activists I've been talking to, Danielle Linder and Debbie Bacigalupi from Siskiyou County. The Klamath River Basin in California is a case study in how the movement destroys a region in order to turn it first to wilderness, then, seemingly inadvertently, to desert.

But first, the general issue exercising the Truth Squad is the passage of Senate bill S. 1867. Its latest title is the National Defense Authorization Act for Fiscal Year 2012, and, despite last-minute provisions, rural activists believe it is a bill that revokes civil liberties.[2] This may be abstract for most Americans. For rural Americans, it is not. They think they have been targeted by federal agencies for elimination by any means necessary, and they have reason to believe that this act will be used against them. When Mr. Obama declared in 2011 that water was not a right, it was a privilege, all hell broke loose. The loss of water rights, property rights, and individual rights, which is an abstraction (so far) in the cities and suburbs, for rural Americans is anything but. Used to the daily incursions by the feds into their lives, land, and businesses, they see it as tyranny. And now that S. 1867 has become law, if they don't comply they are a targeted enemy of the American government.

"'Navigable waters' have been changed to 'waters of the United States,' which means that if you have a puddle in your front yard, the Army Corps of Engineers owns it. The corps now has ownership of every drop of moisture in the land," says Marti, "even the water that falls on your roof."[3]

"And when they own water, they own you. When they own you, you have become their biological resource. And if you don't want to serve as a biological resource, you are a terrorist and can be detained without trial."

Barb further adds, "The White House Rural Council is making rules that the BLM and the EPA will determine whether you are making adequate and beneficial use of water and resource on your property. If they decide you are not, and they *will* decide you are not, they will try to take your land under eminent domain. They will say they are offering you a fair market value for your property."

"If you should say, 'No, I don't want to do this,' the intent is to bring the military in to get you off," finishes Marti.

This is what spooked rural America in the winter of 2012.

The Klamath Basin's fingers reach out to include a collection of towns: Weed, Siskiyou Crest, Happy Camp, Etna, Hilt, and Montague, which are surrounded by three national forests: Klamath, Six Rivers, and Shasta-Trinity. Mount Shasta, 14,179 feet, towers over the valley. Mount Shasta has been repositioned by the movement's usual cloying marketing as a "sacred mountain," and as "the loading ramp for the Ark of the Covenant."

Today, in the early days of 2012, Siskiyou County is almost dead, a case study in an iron-fisted, bureaucracy-driven creation of a no-go zone. Planned as future wilderness where no one lives and no one visits unless it's a government official, the county—as large as Vermont, with 48,000 residents—has been under assault since 2000. Surrounded by the three once-productive national forests, with thousands of ranches and mines and farms, the county is a veritable treasure house of resources that feed, house, and provide the raw materials of city life to millions. Once prosperous, the people here have deep roots, generational roots, some going back to the mid-1840s.

Debbie Bacigalupi is a forty-two-year-old event planner, sommelier, recent MBA, and at present, a full-time activist against the dam removal and the sequestration of ranch land as wetland in Siskiyou County. Her parents' 4,500-acre cattle ranch, Cold Springs, has been under assault for two years. "Oh, they want it bad," she says. "They flew over circling the ranch yesterday, and they patrol along our property line. My father's water master has caught Fish and Game trespassing, and they have turned off our water a couple of times." The Nature Conservancy owns Big Springs ranch down the road, and the Bacigalupis believe TNC has its eye on their property.

"The current trouble started in 2000, when these young men, Craig Tucker, John Bowman, and Glen Spain, moved into the county from the east. Ever since then, we've been under increasingly active attack. They started or ran the local branches of all the organizations that are trying to shut us down, American Rivers [who have forced the removal of 925 dams in the United States and whose stated aim is the removal of all dams], Cal Trout, Trout Unlimited, Klamath Water Users, and the Pacific Coast Federation of Fishermen's Associations. Craig Tucker is now representing the Karuk tribe in its fight to take advantage of the dams being taken out. 'Course the tribes want the dams taken out. The three tribes who are helping to promote the dam removal will receive eighty-seven million dollars and thousands of acres, going forward."[4]

The removal of the four dams on the Klamath River will be the largest dam removal in history. The reason? The coho salmon is endangered and must be protected. But here's the thing: the health of the coho is not supported by the phosphorus and magnesium salt–rich riverbeds of the upper Klamath, and in the 1920s and '30s, before California's industrialization and any warming trend, the records show that, on average, fewer than five hundred fish returned to spawn every year.

"Fish and Game plant coho downstream from our ranch. Even if the dam removal doesn't go forward, to protect the water,

Fish and Game told us that, for the sake of the fish, they are going to impose one-hundred-fifty-foot buffers on either side of our river and streams, which must be fenced, and they are going to call those buffers wetlands. No livestock or people can go into those areas, and Fish and Game will be able to come onto the land to inspect the buffers at any time. If they find a coho fingerling dead on our land, they will charge us twenty-five thousand dollars per fish. How do we know they won't plant dead coho? My parents want to be buried in a place near our springs, and I won't even be able to visit their graves." Debbie struggles with tears, but she recovers fast.

"So what you get is fake 'county' residents coming in here to protect a fish that is not indigenous to the area. The county voted 78.9 percent against dam removal, and all our board of supervisors are against it. But that doesn't stop them; the dams must come out."

Fish and Game told the Bacigalupis that the Klamath Basin is a pilot project. "They will take control of the water in northern California and the ideas and plans move east to take all of the waters in the U.S."

Mr. Obama's declaration that "water is a privilege, not a right" was not a throwaway line meant to finesse a specific political situation; it was a statement that reveals the plan of his recently formed White House Rural Council to remove water rights completely throughout the nation, for if California can't keep its water, no one can. Without the northern Californian waters, held safely for dry-lands irrigation and summer water, almost half of California's $43 billion in agricultural revenue will be decimated. The five counties most affected by the Siskiyou dam removals now provide tens of thousands of jobs and billions of dollars' worth of food production on 240,000 acres. If the dams are removed, within a handful of years those crops and people will be gone.

The two agreements that claim to "solve, mitigate and restore" the issue are the Klamath Basin Economic Restoration Agreement and the Klamath Basin Hydroelectric Settlement. Together they propose $536 million to get rid of the dams and tout another $555 million from private donors to restore the area. Bacigalupi says the $555 million is supposed to come from ENGOs. "Oh, they are all in here working: Patagonia, the Nature Conservancy, and Sierra Club promise us that money for compensation, but as we all know, they get most of their money from the government." That $555 million includes the $87 million for the three tribes who have agreed to agitate for dam removal. If the citizens of Siskiyou sign these agreements and take the money, they sign away any control over the waterways and buffer zones of their counties, putting in place an unelected administrative council that abrogates democratic rights. They also lose their right to sue for damages caused by the dam removals.

And those, it has been proven, can be substantial.

An avalanche of silt and mud and debris will flow from the dams; an estimated twenty million cubic yards of pollutants will flow into the ocean. The dams currently provide clean, cheap, local, renewable hydroelectric power to more than seventy thousand homes and businesses in Oregon and in California, the latter of which has managed its energy so badly that 53 percent of its energy imports is in dirty coal. Irrigation for ranches and farms will dry up, and millions of yards of toxic sludge and sediment will flood the riverbanks, taking an estimated one hundred years to clear through the system. Hexavalent chromium (chromium VI) is a carcinogenic isotope whose compounds are widely used in stainless steel and anticorrosion treatments. Erin Brockovich forced Pacific Gas and Electric to remediate seepage of it into groundwater in Hinkley, California, as documented in the 2000 film *Erin Brockovich*. The chemical was found in the tons of sediment washed down the river from the Rogue River dam removals in southern Oregon in 2009. Allen Ehr, a local environmental specialist and investigator who has been monitoring the environment for the last twenty-five years, found that intestinal cancer rates, as a proportion of cancers in the area, jumped from 17.2 percent to 42.6 percent in two years in the hospital that served the communities directly downriver.[5]

Siskiyou County was struggling long before the dam removal became imminent. California has the second-largest forest in the United States but now imports 80 percent of its wood. Of the 101 million acres in the Klamath in timber, only 7.52 million acres is actively logged.

There used to be twenty-two sawmills in the region; all but two are long gone, thanks to Bill Clinton—bringing poverty, social unrest, and meth addiction, as have all the spotted owl forest shutdowns. Unemployment stands at 25 percent. The valley is now harvesting less than 10 percent of the amount harvested in 1978. To put that number in further perspective, in the Klamath alone, the annual net growth is 130 million board feet. Timber companies harvest, on average, about 45 million board feet a year from the forest at present. Eighty million board feet die. That 80 million board feet could be harvested and sold, but it is not. It is left on the forest floor. And the tinder builds up and up and up. Today, Siskiyou County grows eight times more wood than is harvested.

"And then they wonder why we're having large catastrophic forest fires."

I am sitting in Danielle Linder's comfortable early settler's farmhouse a few miles outside Weed, California. It's hard to get foresters to talk about their forests unless they are retired, at which point they generally are so distressed by the unreason that has shaken their profession that they don't keep up with what is going on. Young foresters are afraid of the movement because they believe if they counter its assertions in public, they'll be out of business. Linder, however, is a decade into the fight, a forester with seven years of university under her belt. She and her forester husband own Jefferson Resource Consulting, which employs one other forester and a wildlife biologist. In 2011 she was named Conservationist of the Year by the California Forestry Association. She speaks in the clipped cadence of an irritated scientist. She cracks only one joke in almost two hours, that having to do with the fact that if "corporate" foresters were rich, she'd be wearing Prada, not Walmart.

It is a commonly held opinion among working country people that the environmental movement has caused the sweeping catastrophic fires that burn through the West from Mexico to Alaska every year, destroying more forests than evil corporations ever could. Siskiyou is a case in point. Fire dynamics for the Siskiyou are fairly straightforward. The Klamath Basin has a lot of lightning-caused fires, especially in the summer and fall. "Fire is influenced by topography, climate, and fuel," says Danielle. "What's changed is the fuel load. You can't manage trees if there are four or five hundred per acre where typically there are sixty to eighty. It's like fireplace dynamics: the more wood you shove in, the more fire you get.

"The species composition is changing too. Our forests developed with fire. The redwoods are a different environment because of the moisture, but you look at the Cascades all the way down to the Sierras, it's a typical Mediterranean climate—hot, dry summers, cool, wet winters. We have summer thunder bumpers [a dry thunderstorm, without rain], so between lightning and the Native Americans who set wildfires, there was a fire every five to thirty years.

"The average fire, then, was a cool-burning underburn, because it was every five to thirty years. It would kill off the brush, kill off a clump of trees. It wasn't a catastrophic, stand-replacing fire; it was a low-intensity fire. Our brush species, by no coincidence, typically have a life span of thirty, forty years. Why live to a hundred years when you're not going to make it? The brush secretes chemicals into the ground which actually promote wildfire. The brush wants to burn, that's how it regenerates."

Throughout forested North America, around World War II, national governments started suppressing wildfire. There were too many people living in the woods, and we didn't want their houses burning down. Stand-replacing fires used to happen in the 1950s, when we started fire suppression, but they just didn't happen with the intensity and frequency they do now. Settlers would clear and burn. Clear-cuts mimic fire. In the clear-cuts, you can stand and fight. "You don't want to be up on some random road in that. A fire will blow right over the top of you," says Linder. "Course, Clinton's Travel Management Plan means the forest roads are being shut down, so you can't get up in the forests anyway to fight fire."

"So we stopped clear-cuts," says Danielle, "and suppressed forest fires, which means more shade and less regeneration of the brush. Overgrowth means no more forest meadows and no more early-succession plant life. The result is old-growth brush up in the trees and an overabundance of trees. Which creates what we call fuel ladders. Once you have vertical fuel ladders, you get catastrophic fires, which is what we have now.

"So, you get all this shade with all this increase of fuels, and you get this white fir in your forest stand. White fir are not drought- and insect-resistant like pine. They have thinner bark, don't like full sun. They burn because they don't self-prune; they look like your Christmas tree. They were here presettlement, but not in the frequency they are now.

"Now you get large areas of white fir dying, because we have droughts every seven to ten years. You have overstocking, which is basically a competition for resources. Think of it this way: trees are like people; if you and I have to share one sandwich a day, we'd probably have fifteen to twenty pounds off us and have a competition for nutrients. If there are too many trees sharing one area, the trees become smaller in diameter; they tend to become shorter and stressed. Stressed trees are like stressed people—they become more vulnerable to disease. Whereas if you thinned out the trees, so there were fewer trees competing for that resource, you'd have bigger, healthier trees. They could fight off insects and disease. And even if you don't 'believe in' climate change, in yet another contradiction, rather than being a carbon sink, the Forest Service thinks that we will be a net carbon emitter, because of the degradation of the forest."

"It's so obvious," I say.

"Yes," she returns, "and we bring state legislators up here, and drive them around, give them a presentation, show them what we do, why we do it, and why it's good for the forest, the county residents, and their tax base. They really get it; it all comes clear to them; then they go back to Sacramento and vote lockstep with the Sierra Club. I call them and ask why, and some don't even bother to avoid me. By now, I've had half a dozen legislators tell me that if they vote against the Sierra Club, they won't win their next election. The Sierra Club will advocate against them, and in the liberal districts the legislators represent, whoever is recommended by the Sierra Club is most likely the winner."

"And the humans?"

"In my son Jack's first-grade class, there are kids living in travel trailers and tents. Tents! In this weather!" It may be California, but there is two feet of snow outside, and the roads are glistening ice.

"I have three sleeping bags in the back of my truck right now because I heard about this family who are sleeping in tents with three children, and the one in Jack's class wanted a sleeping bag for Christmas. I cried. Jack wants a Nintendo. My husband and I decided to buy the family sleeping bags for Christmas. In the last twenty years, we have lost 25 percent of children under eighteen. There are fifteen kids in Jack's class, so now we're talking school consolidation. And 25.4 percent live under the poverty line."

Another weapon wielded by the movement is the Siskiyou Crest Campaign, launched by the Klamath-Siskiyou Wildlands Center (KS Wild), which plans to set aside another six hundred thousand acres of forest in a monument designation. KS Wild's annual budget is $500,000. Danielle, who is director of KARE, the Klamath Association for Resources and the Environment, has an annual budget of $25,000 to fight the monument designation.

"Where are we going to go?" she asks. "You don't have Hollywood or Simi Valley clamoring to come up here and buy us out. If you live here, it's not like you lose your job in Silicon Valley and drive down to San Jose and get a new job. Here you have to uproot and relocate, and a lot of these families are generational.

"In the seventies, when they passed the Endangered Species Act, in this state the conservation movement pushed through the Z'berg-Nejedly Forest Practice Act, which made it a law that to practice forestry in California, it's a minimum of seven years to qualify for a state exam. It's a broad and deep degree, and first, you have to have a BS in forestry from a university accredited through the Society of American Foresters. Once you qualify, they basically hand you a ream of paper and a pencil. The exam is essay form, and you have to get 75 percent or higher on seven different subjects, from environmental law to engineering, to fisheries, forest health, hydrology, wildlife biology, and archeology.

"With that they ensure there are professionals in place to do proper forestry. But they still don't let us work; they don't trust us. The first Timber Harvest Plan I wrote in '93 was thirty-two pages long. Now, thirty-two pages doesn't get you through the introduction. A Timber Harvest Plan is two hundred to three hundred pages now. We're still using rubber-tired skidders; we're still using cable logging. The mechanisms that get the logs out of the forest haven't changed. But the paperwork to prove what you're doing isn't harming the environment has changed rapidly, at the cost to the property owner, as well as society as a whole. Plus, before I had children, I used to get up at three in the morning to supervise tree planting. We planted seven trees for every tree we cut, sometimes thirty, forty thousand a day."

Back at the Truth Squad, they're talking forests, and Danielle describes what they have to do to be able to cut anything on any land, private or public.

"For the last fifteen to twenty years, we've been going out at night to do spotted owl surveys. You have to survey at night within a watershed, and set up a station, so your voice or recording will carry through the watershed, then you have to hoot three times per station for two years in a row before you can harvest trees.

"If we hear them, we have to give them mice, then run through the woods and find their nest."

Emptiness fills the air.

"Shut up," says Marti.

"Now," Danielle continues, "they've doubled the number of surveys you do; we have to visit the station six times for three years instead of two, which has tripled the cost. So if you have timber on your property, you have to have a timber plan done by a registered forestry professional, whether you own five acres or a thousand acres. You hire me, I evaluate your watershed, your land, what trees need to come out, design roads and culverts, evaluate all your watercourses and streams if there are fish, recommend spotted owl surveys, archeological surveys and archeological records check, waiver of waste discharge, erosion control plan. I have to survey for willow flycatcher, fishers, goshawks, osprey, black-backed woodpecker, marbled murrelet, sandhill crane, frogs, salamanders, spotted owls, and threatened and endangered plants, just to name a few; put all of the information in a document; submit it to the Department of Forestry, where the document will be reviewed by the Department of Fish and Game, Water Quality, California Geologic Survey, and the Cal Fire [California Forestry] State Archaeologist. Then they come back to me with first-review questions; they will want a field visit on your property and look at everything. I represent the landowner, so sometimes we disagree. I recommend a culvert, they want a bridge; that's much more cost to the landowner. Once inspection is done, there is a second review, they talk about what they saw, what's in the plan, and you have to answer another set of questions. All through this, you are under review by the public. I have to notify adjacent landowners, and anyone a thousand feet downstream has to be evaluated; also you have to put in a public notice, and anyone can comment on your process. Then after *all* the mitigations are done, the director of Cal Fire will approve your plan, and then you start paying a Fish and Game filing fee and the State Water Resources Board for a waiver of discharge permit; then you get a 1600 [Streambed Alteration] Permit. So unless you have very valuable trees or a very large area, you can't afford to harvest trees. If you live in a neighborhood surrounded by trees and you want to add on to your house, build a barn, there's an abbreviated process, but you still have to hire me, and I have to jump through a lot of hoops, and there are a lot of costs.

"We think 13.8 million cubic yards of sediment will be released downriver by the dam removals, yet when we operate logging along the Klamath River, we can't be within one hundred fifty feet of either side of the creek and we can't get a cup of dirt in the river or state Fish and Game will be on us and we'll be out of business."

"These people make me so sick. I can't stand it," says Barb for the second time.

"What started the KS Wild campaign," I ask, "if so little is being cut?"

"There was a big fish kill in the river in 2002 that the enviros got all up in arms about. They thought it had an industrial source. Nonsense. We think it was a meth lab dumping their chemicals."

"Meth," I say.

"Yes. Meth."

The Green Apparatchiks

Brent and I, Eric and Kimberly, and two trust staff persons sit in a meeting room in the basement of the Community Gospel Chapel for almost forty-five minutes before we realize we have been stood up. A clerk goes back to the trust office to find out what happened and returns to say that no one called to cancel. We wait another half hour in case they got the time wrong, until it becomes clear that the entire sixteen-member Environmental Advisory Committee decided not to come. As we leave, Brent tells me it is one of the ways activists protest an application. They can't find an objection, so they delay until they can.

Another six weeks go by, and only after a half-dozen phone calls and e-mails from me and unrelenting pressure from Kimberly and Eric does the committee finally gather. Briony Penn is the interim head. She is a multitalented forty-something charismatic who gumboot-dances and paints educational watercolors of the animals and foliage of the various "ecosystems" around us. She had a television show for a few years called *Enviro/Mental* and ran unsuccessfully for Parliament. A geographer with a doctorate from Edinburgh, a newspaper columnist, and a university lecturer, she encourages every environmental action restricting use both in the islands and in the region. She flat out terrifies me. Whenever I catch sight of her, I feel like an abused kulak shrinking from the gaze of a party apparatchik.

She slops into the room in an indeterminate outfit of antique waxed green rain jacket, rubber boots, the oldest of khakis, T-shirt, and pilled fleece vest, carrying a giant file folder. I am pretty sure she has not brushed her hair this morning; it is caught up in a ponytail, which has partly pulled out of its binding. She sits down to a chorus of greetings and is immediately asked to become the permanent chair.

She smiles and refuses—"so busy," she says, her speech revealing a very slight English accent. Briony is a born patrician who has become a vigorous anticapitalist. As she paws through the welter of paper in front of her, Kimberly takes pity and walks her through the procedure she must follow. Eric sits silent, staring at the screen of his laptop. Brent sits beside me, stalwart, radiating virtue and confidence. I resist an impulse to grab his hand and hold on.

Briony asks the committee members to introduce themselves. Each has a doctorate or an advanced degree in some branch of the life sciences. One recently arrived American is an archeologist, her purpose on the committee being to speak for First Nations burial sites and other indigenous points of interest. Each of them may have a specific objection to my application, based on their superior knowledge of earth, air, water, and endangered creatures.

There's a rattle at the doorway, and the final committee member, Kathy Reimer, comes in, carrying five huge rolls of maps. She catches Brent and me in a brief sideways glance and manages to convey encouragement. With a slight sigh, she dumps all the maps on a side table and takes her chair. The mood in the room shifts a bit. Despite all the biologists, botanists, geographers, and other experts on the committee, no one but Kathy actually gets out onto the island to restore it. She knows the island in ways no one else fathoms; she is that necessary thing, a dirty-hands restorer. The maps are there to bolster her argument that my subdivision will enhance the island's salmon habitat.

Briony roasts me over a slow fire, muddling through a welter of committee rules, moving with inexorability toward my application, the only item on the agenda.

Quarreling is a way of life in the life sciences, as it should be. Jackie Booth, a marine biologist, challenges Kathy's desire to dig a pond for salmon. It will attract mosquitoes, she says, and there is dubious benefit there. No salmon will ever appear in that creek, she suggests. Kathy—the implication is slight, but present—is feathering her own nest.

This is a familiar argument. Most activists believe that land must be left alone to recover, in a passive restoration or "natural regulation," which most federal and state agencies, one way or another, have defaulted to over the last forty years. Introduced in the 1960s to Yellowstone, natural regulation has more failures to its credit than successes, but the idea is so persuasive—probably because urbanites, whether transplanted or not, "believe" in letting nature be nature so fervently that the idea that the earth is man's garden rather than wilderness has lost all currency.

Others believe that only by actively fighting invasive species and aggressively reintroducing almost extirpated species can an "ecosystem" recover. Fifteen years ago, when I first walked my lower meadow, it was a soggy mess; I sank into it up to my ankles. It was wetland. Now it is dry, the aquifer having retreated. Some of that retreat is due to wells pulling water from the springs above us on the mountain. And until the last two years, summers had defaulted to a weather pattern not seen on the coast since the 1920s and '30s. My father used to reminisce about his childhood summers with ninety days of full-on sun. We'd mock him mercilessly; our summers in the 1970s were all too often wet and miserable. But as Alston Chase suggests, everything is always changing, so you manage landscape for what you want. Do you want wetland species and salmon swimming upstream? Building a pond encourages wetland species to return. Introducing brown trout into the pond brings life to it; everything percolates and shifts a little bit. And maybe, just maybe, one day, a salmon will swim up that stream and spawn, and her fry will summer in the deep holes. It sounds good, doesn't it? It did to me too.

After two hours of tortuous deliberation during which I nearly lose consciousness from worry, they give me a conditional pass, because, as Briony sighs, closing her folder and calling for a vote, "I can't find anything wrong with this application." They have scheduled a site visit, at which Kathy will fight for her pond, after which they may approve the application for the second and third bylaw readings. The fourth is supposed to be a mere formality, just a completion of paperwork. The next step is the critical one, the one that could, if I haven't done my homework, call out the mob.

Kathy and I return to her ramshackle office over the strip mall, and she pulls map after map out of her files to show me the likely success of our project. She is babbling and confused, as if her back is to the wall. I realize she is just as afraid of them as I am. She is right to be afraid. There was little doubt that the strict constructionists on that committee want Kathy's work to stop, so that the forests she works in can return to the tangled mess they supposedly were before the plague of white men hit the continent.

CHAPTER EIGHT

The Real Green Economy

In California, nine hundred miles south of me, Crescent City in Del Norte County is as close as I can come to a twin to my Gulf Islands. Del Norte is the gateway to Redwood National Park—home of the iconic trees, the threatened loss of which inspired the fund-raising of whole generations of San Franciscan socialites.

Del Norte's appeal, therefore, is much like that of my Gulf Islands, a supposed magnet for tourists willing to pay to be awed by natural beauty, with a similar population of twenty-nine thousand. The tourism business agrees, apparently; the "city" was displayed on the December 2009 cover of *National Geographic* as the archetypal northern California town. That same year, ForbesTraveler.com trumpeted Crescent City's "pristine parks, miles of beautiful and quiet beaches for swimming, miles of rivers for rafting, and world-class fishing," making it one of "the prettiest small towns in America." Along Redwood Drive, which runs close by the ocean, hotel chains have erected their representatives: Econo Lodge, Best Western, Quality Inn, and the poshest, Hampton Inn and Suites. I stay at the Lighthouse Inn, where the staff put me in mind of the cast of *Glee*, with an added whiff of desperation, and the decor is delightfully idiosyncratic. The desperation is because the promised tourism hasn't turned up.

Del Norte County is instructive because the board of supervisors has commissioned, over the past few years, an entire raft of number-rich audits of the effects of the environmental movement on the economy and the people of Del Norte. Their work puts paid to any idea that green means wealth for anyone but those paid by green to advance green. On the other side of the ledger? Bankruptcy, crime, poverty, and hunger.

Crescent City is beautiful, as long as you keep your eyes firmly fixed on the ocean, which is rough and wild and exciting. The town itself? *Dilapidated* would be the kind word. *Deserted* is another that springs to mind.

It has been a long time since anyone did much exterior maintenance on the buildings that house the town's various businesses. Besides, as the population has dropped, calls on those businesses have dropped too. There are no tourist attractions other than a giant lacquered redwood on a stand in a run-down oceanside park. You have to be fit and young, with money and a guide, to enjoy any of the vigorous tourist activities touted by *Forbes* and *National Geographic*. Restaurants? None that would attract any reasonably well-heeled tourist.

But the Smith River, a "Wild and Scenic River," is one of the cleanest in the world, and the air measured as the purest in California. In its fishing heyday, Crescent City's harbor was the most productive in northern California, but today its catch has been reduced by one-half.[1] For most of its existence, Del Norte was prosperous, independent, and largely working class. But today, despite the PR bluster, Crescent City is a shaken town, and Del Norte County has endured catastrophic change.

And, of course, because it's California, it's also broke.

Which may explain why legal counsel to the county commission, Dohn Henion, punctuates his story with a laugh that sounds—after you hear it for the tenth time—more like a sob.

"County commissioners and their lawyers see everything," explains Henion. "We are the only accessible authority in town, so when people are in trouble, they head to us."

I'd head for Henion too, if I were in trouble. At fifty-five, he is as beefy and muscular as a football player. He looks as if he were born in a navy blazer, his quiet seriousness emphasized by the brass buttons, the black brogues, the gray flannels, the utilitarian uniform of the grown-up. Henion has had his own practice since 1981, and at first he dealt mostly with property law. Then he signed on as a city attorney, and subsequently city prosecutor; he enjoys trials just that much. He represented police unions and then represented public entities against them. And for ten years, he was the prosecutor for Indian affairs on the Hooper Reservation. It was there he hit a career-making case that he won at the U.S. Supreme Court level. "Went on for seven years," he says today. "Still haven't been paid. Don't think I even billed them."

Henion has pleaded both sides of most every issue that comes before him as counsel to the county commissioners of Del Norte. And there are many issues. On the day we met, he had worked fourteen-hour days for twenty-five days in a row, and that was by no means unusual. The business of Del Norte County these days is the business of environmental government, and it is fiendishly complex. Ever since Redwood National Park was established in 1968, then expanded in 1978, then expanded again, Henion wakes up every morning wondering which of the nineteen agencies[2] with an environmental mandate, which control every aspect of life in his county, will drop a bomb on his desk today. And where he's going to get the money to clean up the human debris that will inevitably result.

Once upon a time, Del Norte didn't need help. It was a largely ignored logging and fishing county. The sea is tough, the forests are big, and life was lived mostly outside. Even though 22 percent dropped out of high school and one-third couldn't read, the smart kids, the ones who wanted it badly enough, were sent to college. Retirements were paid for. There was security in this town for ordinary people. The poverty rate was lower than the California average.

The story that follows is best told through numbers.*

The 132,000-acre Redwood National Park was created in 1968. Immediate job losses in the forestry sector and among those servicing the forestry sector totaled 2,757. In 1978, the park was expanded, and this time job losses totaled 4,218. More than 25 percent of Del Norte County's population was now out of work, and these were family-wage jobs, not McJobs. Every year, more and more acreage was slid into the park, and more jobs vanished. For instance, when the 24,753-acre Mill Creek property was added to the park in 1990, another 900 jobs were lost. State parks were folded into the national park in 1994.[3] In another four years, the poverty rate in Del Norte County had reached 22.9 percent, up from 5 percent before the park was founded. Property taxes, on the other hand, were way down. Tax loss on the Mill Creek property alone was $6.4 million a year. Today, 31 percent of Del Norte County's children live below the poverty line.

Then there was the tourism chimera. In reality, 90 percent of the park is inaccessible to the average tourist. Funds to equip the park with campgrounds, picnic tables, and other facilities never turned up. Instead of a predicted 1.6 million visitor days by 1983, in fact, over the first fifteen years of the park, there was a grand total of thirty-nine thousand visitor days.[4] Tourism days reached less than 5 percent of the visits projected by the Sierra Club's much-cited Arthur D. Little study, and less than 4 percent of the National Park Service estimate promised to Congress at the

* Many of these studies are archived on www.elizabethnickson.com.

time of the creation of the park. And that tourism is sketchy at best. Rather than an average of twelve hours spent by each tourist in the park and region, which had been projected by the Sierra Club's study and the NPS, most people drive through the park in fifty minutes and bugger off up or down the coast to have some fun. Fewer than 10 percent get out of their cars to dance, or even picnic, under the massive redwoods.

The cost of the park, as promised to the Senate in 1968, would be under $92 million. Once all the acquisitions of logging tracts and state forests were completed, as of 1995, Redwood National Park is the most expensive of the national parks, at $1 billion.[5]

But if the unemployment and poverty rates in Del Norte County have quadrupled, the education rate has shot up too. Suddenly more than a third of its residents have a college education. Thirty-three percent of the county is now employed by those same nineteen agencies, double the percentage employed by government in the rest of California. Environmental government jobs pay substantially more than private-sector jobs, so those 33 percent earn 50 percent of all wages, more than three times the percentage earned by government workers in the rest of California. However, the median income, despite these flush enviro jobs, has dropped from $55,000 to $32,000. Two-thirds of the people exist on subsistence wages, in thrall to their green overlords.

"The park expansion turned us into a welfare county," says County Supervisor Gerry Anderson, a mechanic. "There's a long collective memory among the locals of what it used to be like." Many turned to hunting and fishing to supplement family diets, but incrementally, over long months and years, the best fishing sites were closed off, and access to the forests for hunting was forbidden.

Henion and the county supervisors ceaselessly look for ways to employ their workforce. "Our prime opposition is the California Coastal Commission," says Henion. "They supervise everything from the coast all the way to the base of the mountains, and everything, to them, is environmentally sensitive habitat. They forbid vacation rentals and time-shares and now want us to prevent people from staying in a hotel longer than two weeks. On a fifteen-hundred-house subdivision they refused to let go forward—people have been paying property taxes on that vacant land since 1964, waiting to be able to build—they refused to permit any economic activity. An RV park? No. Yurts for campers? No. Trails? No. A parking lot? No. We need a permit to turn over a shovelful of sand."

After the recitation of facts, the look in Henion's eyes is tortured. He is, for the first time in his life, facing utter defeat. And defeat in service of caring for the weakest in his county. I imagine that Henion, even from his earliest years, was no innocent. He is an equanimous sort of guy. Nothing rocked his world; crap happened, after all, and a lot of it, and though not particularly religious, like most of us, he always believed that the goodwill of God's people generally prevailed. But absurdity piles on absurdity. In 2000, the California Coastal Commission informed the county that no agricultural lands were to be subdivided or developed, thus pitching the farming community into near bankruptcy. Then California ordered the county's state-of-the-art, triple-lined landfill shut down and the refuse shipped to Oregon. The Army Corps of Engineers refused permission to build the necessary transfer station because there was an endogenous wetland plant species in the ditches at the side of the road.

Henion watched the California Air Resources Board collect his county's air during the cleanest time of the year and subsequently force the rest of California to conform to that preposterously high standard. And when the feds insisted the county's simple airport measure up to new terrorist-era standards, he was again held for ransom by the Coastal Commission, which now insists that passengers walk the length of two football fields to their planes, through a supposedly endangered Sitka spruce forest, the trees as common as mud everywhere on the coast. With that and other refinements insisted upon by the army of green bureaucrats who feed on Del Norte, the airport building's cost was doubled.

Henion can't pay any of the county's bills. Only 22.4 percent of Del Norte is private property now, and from that he and his commissioners draw property-tax revenue to run county services. The county is supposed to be paid PILT,[6] but that money hasn't shown up for years. "When the Forest Service offers a grant to allow people to clear out the underbrush and deadfall to prevent forest fires," he says, "I beg for the money and the temporary employment. Men come to my office with tears in their eyes. They can't feed their families."[7]

When Dohn Henion and his commissioners wake up every morning, they wish Del Norte wasn't so beautiful, because then perhaps they wouldn't be coping with so much human misery.

I grew up as a city person. My father moved us to the country when I was six, but I remember what we were told: this is the country; the beat of time slows; give it a year before you run back to be stimulated by city life. And the people who live here? They know more than you; they have encyclopedic knowledge, a depth of knowing both ancient and modern. They understand the rhythm of earth and sky, lake and season. You do not. Shut up, at least until you learn to respect them.

That's not how we roll anymore. Urban activists moved into communities like Crescent City and with a prodigious energy set about reconfiguring the place, even before the spotted owl was listed.

Ron Arnold, a former executive director of the Sierra Club and founder of the unfairly maligned Wise Use[8] movement, has spent the last twenty years researching the cooperation among foundations, ENGOs, individual activists, and activist federal employees.

Arnold calls it a bewildering array, his felicitous phrase hiding the ferocious complexity of his work. Using social network research in a series of exhaustively documented books, Arnold proves that thousands of activist members of advocacy groups are employed by federal agencies in positions that give them opportunity to exercise agenda-driven "undue influence" over goods-production decisions applied in rural areas. Put plainly, by the early 1990s, according to Arnold, the federal agencies—the Forest Service, the Fish and Wildlife Service, the Bureau of Land Management—and many of the equivalent state agencies were riddled with activists.

Arnold describes how networks of environmental organizations coordinate to systematically target specific rural communities for economic dismantling. They are steered by grants from private foundations, which have become prescriptive rather than responsive. Foundations, says Arnold—and his assertions are backed up by peer-reviewed research[9]—are now trying to create new economic paradigms. Foundation employees, agenda-driven bureaucrats, and ENGOs operate without any oversight by anyone in the larger culture and do not have either accountability or liability for the dislocations they have caused.

Private foundations design the programs—say, the introduction of the grizzly and timber wolf into the forest and ranching communities of the Cascades and Rockies—and select an array of activists and decide on specific tactics. Finally, their PR outfits paper the public square with reports of the success of the movement's work. The Hewlett Foundation runs a glossy communications program for activist NGO and foundation employees, which teaches them how to sell their message; as a result the movement speaks with a single, focused, scrubbed, and polished voice.[10]

Many researchers have attempted to track the money sloshing through the environmental movement's coffers, but the numbers are so large, they very quickly cease to hold meaning. In order to grasp its magnitude, it is useful to sketch a kind of pyramid starting with the United Nations and the European Union at the top accompanied by the International Union for Conservation of Nature (IUCN) and the various UN environmental programs, along with the august foundations of American capitalism and associated ENGOs; together they devise the direction of the movement. It was in 1992, under the leadership of the Rockefeller Foundation, which started the umbrella organization, the Environmental Grantmakers Association, that larger foundations became prescriptive in their giving, and twenty years later, the EGA and its partners design campaigns and masterfully steer the culture in any direction they choose, again with little liability for the change they bring. Any campaign is coordinated over many organizations, from the big dogs like Sierra Club and TNC, down to regional groups like Cal Trout, and to local individual paid activists who move into the county and join or start lobbying organizations. Sniper fire is contributed by litigating environmental groups like the fearsome Center for Biological Diversity or the Sierra Club's Earthjustice.

For years, shocked researchers measured the might of these activist foundations, coming up with figures that hovered around $3.5 billion for the top tier alone. This is no doubt still the case, but the EGA has closed access to its own measuring, in response, one imagines, to furious criticism from rural people whose lives are most affected. We know from a study performed by the Urban Institute with the cooperation of the EGA in 2008 that as of 2005, there were 26,500 environmental groups, and by using median income and equivalents, from the data provided, those 26,500 groups spent about $9.7 billion a year;[11] the top fifty alone spent $4.98 billion. Seven years later, it is almost certain that money has grown, but the updated study is embargoed unless you are a member of the EGA and give more than $50,000 a year to environmental organizations.[12] Nevertheless, that money is just the proverbial tip of the iceberg.

These foundations and their ENGOs, all with extraordinary access to the public purse, act as a massive lobby in Washington and in every state capital, regional government, or county, working with the U.S. government to prosecute the environmental agenda in local jurisdictions where, after all the conferences and posturing, the money does its work.

The larger ENGOs are partner members of the UN and work with the UN Environmental Program (UNEP), which acts as Big Daddy or overlord in the system. The UNEP spends $260 million[13] devising the general direction of the movement[14] every single year. "Cooperation with civil society and youth" is a principal mandate, which sounds anodyne until you are the subject of a masterful campaign created by "youth" and "civil society" devoted to adding another 15 percent to your costs, after which you will be out of business. Equally, on the face of it, it is impossible to quarrel with the notion that coral reefs and great apes should be preserved and studied, or that a billion trees should be planted. Except that, almost certainly, the desired result around the reefs, apes, and trees will be a curtailing of economic activity as in Crescent City—apes, trees, or coral reefs be damned. The IUCN, which is made up of the principal ENGOs and most nations' national land-use agencies, has spent over the last decade $100 million a year ($153 million in 2012)[15] devising the "science" upon which restrictions are based.

The European Union, without doubt the UN's senior partner in imposing strict environmental controls, spends another $260 million a year granting environmental groups the funds to agitate and terrify Europeans. And that doesn't count the flow of funds to environmental groups from each national government in the EU, or the provincial and local governments within them. Nor does it count the funding from European foundations and corporations, all of which are subject to substantial annual green demands.[16]

The drumbeat of doom from all that money is deafening. Little wonder that in July 2011, when EU Climate Action Commissioner Connie Hedegaard announced a threefold increase in her budget, she was able to say:

> *If we ask the Europeans, they are in no doubt: a recent survey shows that almost nine out of ten Europeans favor allocating more EU funding for environmental and climate-related activities.*[17]

The lobbying of government and civil society by the UNEP, the IUCN, the EU, and ENGOs has been successful beyond imagining. Since the Rio summit in 1992, Western nations have contributed about $68 billion annually to developing countries in "sustainable development funds," meaning that most development in the developing world is now sustainable development. To get an idea of proportion, humanitarian aid stands between 5 and 7 percent of bilateral and multilateral sustainable development aid.[18] The U.S. federal and state governments spend about $44.5 billion a year[19] on environmental cleanup, instituting sustainable practices, conserving land, and generally prosecuting the agenda devised in partnership with the UNEP and ENGOs; after all, federal land agencies are members of the IUCN and collaborate as full partners. Some of that money bleeds out to activist groups. We know, for example, that over the past ten years, the EPA—whose 2010 budget shot up 30 percent—has granted $3.8 billion to groups[20] advocating for environmental justice. What this looks like in the press or on television is genuine citizen outrage. To the contrary, no matter where you live, it is almost certainly your government paying people (useful idiots or not) at arm's length, to scare you into believing that you need more government.

Ten percent of Mr. Obama's stimulus, or $78.6 billion, was dedicated to clean energy projects, energy efficiency initiatives, and green transportation.[21] But it is business that surprises the most. Rightly or not, the business sector appears to have been captured by the movement, and in the United States and Canada, by 2014, sustainable business expenditure by corporations will dwarf spending by any other sector of the economy, standing at more than $70 billion.[22] In the EU, that figure, assuming a reversal of Europe's current plight, will be up around $97.98 billion.[23]

Factor in the work of universities. Fifty-three environmental law journals are published in the United States; for instance, both Stanford and Harvard publish their own. Environmental law faculties ceaselessly devise new methodologies for environmental protection; this would be expensive if the NGOs were required to provide that research themselves. Biology has been almost entirely captured by conservation biologists, and universities publish conservation biology journals and studies in an unending river of agenda-driven, politicized science.

Every time you see a story in the newspaper that some worthy-sounding group you've never heard of before has announced a new "study" showing that economic activity must be sharply curtailed or disaster will occur (or far less frequently, has occurred), consider that you paid for that study and that it probably has not received any scrutiny from a dispassionate, independent scientific body. In-house peer review or toss-it-over-the-cubicle-partition review is standard practice. But somewhere some poor sap who owns a business has to find another ten, twenty, or hundred thousand dollars to conform to the new regulation this study has triggered, thereby crippling his profit margin and too often putting him out of business.

My jurisdiction, counting the money that funds our local conservancies, sustainability institutes, and conservation government, spends almost $9 million every year* managing and improving the environment and land use. Our population stands at twenty-five thousand people, so that much money entirely dominates our culture. Salt Spring, like a few hundred other similar communities, sees itself as a model for every township and county in the world. If the movement has its way, we *are* the future, and it is not an enviable future; it is poor, heavily restricted, with population, services, and economy all in decline.

The money on the other side? Virtually none. American Stewards of Liberty, based in Taylor, Texas, is the longest-standing activist group for country people. It has a budget of less than $350,000 a year, but their office, with a full-time staff of four, is the richest and most effective advocate for farmers, ranchers, and other country people in the whole of the United States. Oil

* We have a half-dozen land trusts and conservancies operating in our area. The Salt Spring Conservancy, for a population of 9,500, spends almost $1 million every year on "outreach," "education," and land acquisition. The Islands Trust budget stands at $6.9 million, the Land Trust Alliance, $150,000. The conservancies on other islands, the Gulf Islands Alliance, and the Salt Spring Sustainability Institute all work toward preservation and sustainability. I don't include the substantial amount spent by the Ministry of the Environment, or any of the regional NGOs that seek to modify our behavior and economy. U.S. foundations have spent $300 million over the past ten years, mostly in British Columbia and Alberta, promoting their no-use or limited-use agenda. So many questions have been raised about Tides Canada's political activities, that it is under audit for the second time in three years.

money? Only on the other side.* Of the ten thousand property-rights, ranchers-rights, and ratepayers groups and so on, very few have budgets greater than $50,000 a year.[24]

Smart regulation is developed for most industries based on applied science. The industries demand that it be targeted and precise, with detailed metrics; they know that if they do not weigh in on the process and make sure the regulation is smart, they will be forced out of business. You don't want to fly to Europe on a plane for which the metrics determining the wingspan were kind of exact. Most industry participants would say that regulation protects their businesses and serves them well. But when land is public, or if it can be made quasi-public with so many regulations attached that most property rights are removed, people can't afford to fight, having lost the better part of their wealth. So ideally for the movement, regulations should be almost infinite in reach and so imprecise as to be interpreted in a dozen ways. All that's needed to force these regulations is sufficient documentation of collapse, supported by books, movies, documentaries, museum exhibits, cartoons, newspaper and magazine stories, the devotion of fervid columnists, et cetera, ad infinitum, to convince the public that "something must be done." This "feeling" is backed by slightly more substantive hundred-page full-color glossy PDFs aimed at policy makers and politicians. If any kick up a fuss, the lead pony on whatever campaign it is proffers studies from a dozen "independent" scientists. And with a lawsuit waiting in the wings if the policy makers do not submit, presto, you've moved that land into the public sphere. With their billions to spend, the dominant ENGOs are in sandbox time—it's so easy, it's fun.

* The Energy Foundation began as a project of the Trust for Public Land with a $21 million contribution by Pew. The Energy Foundation wound up business in 2008, but for more than a decade, the seven largest foundation funders (among them Pew, MacArthur, Packard, Hewlett, and Rockefeller) of green activists participated in the Energy Fund, which manages their combined investments of billions in oil and gas companies. Source: The Energy Foundation IRS 990, 2008, and http://www.undueinfluence.com/energy_foundation.htm.

After the land's useful value is reduced, regulation can do whatever it likes. And regulation likes to grow like kudzu, especially among the new green bureaucrats, each of whom needs to make good before the next review. After all, tens of thousands of hopefuls paw and snort in the holding pen, fielded by the environmental law and science departments of colleges Ivy and state, community and technical, looking to get in on the richest public-servant money train in history.

In 2001, Tom Knudson, a reporter from the *Sacramento Bee*, wrote a five-part Pulitzer-winning series on the modern movement. One of those pieces, "Litigation Central: A Flood of Costly Lawsuits Raises Questions about Motive," was the first, and regrettably only, mainstream takeout of the movement's activity in the courts.

Here's how it works. A local activist group[25] like, say, the Blue Mountain Native Forest Alliance, will hear of a cut block about to be leased to a logging company. It contacts the Bullitt Foundation (the KING 5 TV syndication fortune, which bankrolled the Oprah franchise) in Seattle, which funds the activist group's commissioning of research (generally into "threatened" pairs of breeding owls in the area), as well as helping set up the community actions and protests that follow. Bullitt will then pass the baton to one of its colleagues to fund a suit against the proposed opening of the cut block. There are many environmental law firms; the Sierra Club's litigation arm is called Earthjustice and is the big dog in the fight.

Earthjustice will then take the Forest Service to court. There is no upper limit on the cost of litigating by either the Forest Service or Earthjustice. The incentive to settle, therefore, is nil, so the litigation can wend on for years, during which the forest is untouched, so Earthjustice is happy. Mounting bills are paid by the unwitting taxpayer either through taxes or through donations to an umbrella organization like the Sierra Club or the Wilderness Society or any of a thousand organizations as well as by higher prices on the commodity. In the meantime, the jobs promised by that cut block or mine or rangeland permit vanish. The defendant had to post a $100,000 bond to apply for that cut block; if he is a small-business owner, his money languishes in limbo. Only large corporations have the money and muscle for that. Therefore the small businessman is driven out. In general, however, even the big corporations don't bother to get in the National Forest game anymore. The healthiest productive forests in the United States are privately held by insurance companies and called "timcos" (timber companies); timco managers work hard to stay out of the public eye, because the movement will litigate private forest owners if they think it will advance the cause. Most large corporations instead donate large sums to the Nature Conservancy, Sierra Club, et cetera, hoping to buy cover or at least put themselves in a stronger negotiating position if they ever, God forbid, try to exercise free enterprise in America's vast rural inland empire.

Alternatively, a conservation group just buys the forest and shuts it down. I had listened to Denis Hayes, the director of Bullitt and a cofounder of Earth Day, tell me excitedly about how enthusiastic his donors had been when they saved the Loomis Forest—close to Ferry County, Washington—from logging in the late 1990s. "We raised sixteen million dollars in just a few weeks to preserve twenty-five thousand acres of forest, and not one of the donors will ever go there," he exulted, a win-win-win, as far as he was concerned.[26] There are no statistics available on what happens to lands preserved by private interests, but the evidence of degradation on public lands is so overwhelming that one must assume that private "saved" forests are dying too. After all, it took only seven years for the Loomis to explode in fire, in August 2006.[27]

Holly Fretwell claims that for every dollar used to buy land, another seventeen must be spent every year to caretake it, using controlled burns, eliminating exotic species, maintaining access in case of fire, and removing woody debris.[28] Hayes will take no interest in raising $102 million every year to caretake the Loomis Forest,[29] not that it exists anymore. In economic terms, this is the opposite of a virtuous circle.

The Forest Service, which used to make money for the nation by selling its timber, now loses, in timber sales alone, half of every dollar. Much of that money goes for litigation; as Fretwell points out, the Forest Service is paralyzed by litigation, appeals, and a confusion of science and politics that results in something called analysis paralysis. Half of the money that the Forest Service used to receive was used on restoration. No more. Since timber receipts crashed, management budgets and staff have been reduced. Rational forest planning and management by professionals is almost impossible.

Evil may be too strong a word for us modernists to use comfortably, but what else do you call an idea that ruins everything it touches?

You Have the Right to Pay Property Taxes

I have adopted a stolid, rocklike refusal to disappear, which is surprising, because disappearing is something I do rather well. Even more surprising, I will call anyone about anything at any time. My shyness is as residual as a prehensile tail. Furthermore, rather than flee from people who appear to dislike me, I start wagging my tail and skid over to them smiling. I reason that it is hard to sit in a room with someone you hate for eight hours without softening toward them just a little.

So I go to meetings and sit there, modest, no attitude of confrontation or objection. This approach, as I have gone around the various interest groups on the island and made my case, might be working. The price was high, for my pride at least. I have petitioned my enemies and solicited the help of community leaders. For an entire month, I waited till suppertime, pulled on rubber boots, and went up and down the two public roads bordering my property, with my map, my plan, and my list of the ways in which my plan conforms to the community's plan and even to its faintest wishes. A couple of neighbors told me later that activists had come into the neighborhood to stir up anger about my proposal, but I, shamefaced and pleading, had got there first.

I sat in meetings where I was distinctly unwelcome, and I have made an enormous effort to understand everyone's point of view, despite reason shrieking at the base of my lizard brain, where I have pushed and imprisoned it. What I have done is accelerate a process that routinely takes three to five years, by obeying every green *diktat* and every green wish, making my subdivision conform.

A week before my public hearing, I get on the phone and beg my friends to come to the public hearing and the subsequent meeting, where my second and third rezoning bylaw readings are on the agenda. And if necessary, could they please stand up and defend my application? It is not, I argue, because there is anything wrong with the application. But you never know. At Janet Unger's public meeting, the "community" ganged up on her, and the furious objections of her neighbors and the conservancy put her in the hospital.

I trail my dismal self to the hall and immediately meet Susan Russell, who sits beside me, sunny, confident, powerful, knitting. Her husband, Denis, an engineering professor, sits beside her, willing to refute any objection about water or siting. A dozen other friends attend, smiling but jittery, pretending confidence.

Gary Holman, our regional director and a strongman in the shut-down-everything movement, tries to put a public walking trail through my property, then complains that I will make "too much money" selling my house. As a final blow, he insists that my water rights be removed.

Other than Gary, no activists. Not one.

Brent, tall and bulky, stands up and describes the subdivision in some detail. Everyone has heard it all several times before. Trust chair David Essig asks an excruciating three times if anyone in the audience has any objection whatsoever, and no one stirs. The absence is crashing. My bylaw slides into regulation. I fall into bed and sleep for an entire weekend.

The month after my second and third bylaw readings go through, I attempt to get the fourth and final reading onto the agenda.

"Complaints have been made," I am informed by the trust secretary.

I go to the meeting. And in the public information section, David Essig, the head of the Islands Trust, makes a special announcement.

"In response to the complaints received by the trust office about the rezoning of Elizabeth Nickson's property, I have gone into the trust files and found that, in ten years, only one in one thousand applications like that of Ms. Nickson has been approved. This single success met all the requirements and exceeded them, and there was no public protest. The passing of this one additional rezoning, which allows a twenty-eight-acre parcel to be divided roughly in half, does not mean that rampant development has come to the island."

With this seeming vote of approval, I go to the trust office after the meeting. And meet a brick wall. Apparently if the island activists can't stop me, they can slow me down. John Gauld, the regional planner, has "decided," he informs me, that my rezoning is going through "too fast" and must be delayed, in fairness to other applicants. I can't apply for subdivision until final reading of my bylaw. And I can't list the house until I apply for subdivision.

As a result, I miss the summer selling season and end up paying interest for an additional year on the loans I have taken to put through the application. If we had not been nearing the crest of a once-in-a-lifetime property boom, I would have been ruined by this one decision alone.

I wait seven months for the fourth bylaw reading, which is pure formality; there is no question that it will pass.

I continue to mishandle myself through the bureaucracy. I become so emotional on the phone to Land Titles in Victoria that a woman clerk actually puts me on a time-out. When I go to Land Titles to register the many things that must be registered and deregister things that must be deregistered, I watch messengers from various legal firms line up to hand in their paperwork. Suburban subdivision after subdivision goes through.

The average title on a piece of land meant for one house is two pages long. My title, with its various attachments, is 126 pages long.

Between Kathy and her salmon project and the restrictive Islands Trust covenant on the ravine and forest, I have lost all but 4 acres of the original 28. I still pay taxes on the 16.5 acres I supposedly have left, but I'm lucky I am allowed to walk on it. And if they want to stop me from walking on it, given the fuzzy metrics of conservation science, they can, and someday they will probably try.

CHAPTER NINE

It's Not About the Spotted Owl

I am standing on the flatbed of a three-quarter-ton pickup with Kathy McKay of the K Diamond K Ranch in Republic, Washington, hanging on to a bale of straw as the truck rocks its way down a steep incline into a vast field. It is snowing and the snow is already two feet deep. As we lurch and grind, about a hundred horses spot us, turn, and as if animated by a single puppet master, start to run toward us. They are backlit by snow-covered trees ranked up the snow-covered mountain.

For the next ninety minutes, we peel six-inch layers of hay off the bales and kick them in pieces into a gaggle of horses, then jerk on to the next stomping, nickering group. A slip on the mud and slush and I'd be under the feet of six or seven dancing hungry horses. But the exhilaration is inexpressible, and not for the first time I envy the people who live out here, who live like this, working outside every day no matter the weather, using their muscles and sinew for a purpose other than "health" or longevity. There is a sense here that there is no place else. For me, Ferry County, Washington, has a kind of limerence—I've known about its drama for years, and seeing its beauty, I understand the dedication of those who are so beaten, so thoroughly thrashed, outmatched, and ruined. It is as if giant psychotic five-year-olds had moved into their county, ripped out its industry, pulled up the train tracks, broke the weirs and dams, introduced predators to kill cattle and horses, and methodically ruined family after family, ranch after ranch, forest after forest. And then left, delighted at their "progress," never to return.

"We're dying here," says Republic mayor Shirley Couse. She has a cold today, so she sniffles through our meeting. She is a volunteer mayor. At first she stepped into the post when someone fell sick, and since then no one has run against her. There's nothing fun about managing decline. She ticks off her problems, then adds, "The only thing that's saving us is the gold mine that was recently reopened.[1] And even with it, we are a welfare county."

Ferry County is the poorest rural county in the state and is the U.S. county most affected by the actions of environmental activists. Once rich, with a high median income, now desperate, still it shimmers with gold, and an occasional fantasist like me can see the glitter underneath the snow and trees, the wide flat rivers and strip malls, junkyards, and gas stations. Gold founded Ferry County, and surveyors claim the region holds all twenty-nine minerals named in the Bible. Ferry and its neighbors—Stevens, Colville, Okanogan, all the counties in the Columbia basin—together form a lost fairyland of dense forest, white-capped mountains, narrow valleys, rivers, creeks, and wetlands—like Lothlórien, the Land of the Valley of Singing Gold from *The Lord of the Rings*.

The action that started the ruination of Ferry County is the most stunning success of the modern environmental movement, the northern spotted owl campaign in the 1990s, which shut down 90 percent of the productive forests of the American West. It required only a few months of marching, political pressure, direct actions (sometimes called ecoterrorism), and a typical Clintonesque deal, which drew off some of the Left's fire for his ratification of the North American Free Trade Agreement

(NAFTA), but embedded in that campaign lies the corruption at the heart of the modern movement. Andy Stahl, then resource analyst with the Sierra Club Legal Defense Fund, declared: "Thank goodness the spotted owl evolved in the Northwest, for if it hadn't, we'd have to genetically engineer it. It's the perfect species for use as a surrogate."

And a surrogate it was. The reason the bird was so convenient was that it ranged over an extraordinarily large home area, each breeding pair apparently depending on thousands of acres of old-growth forest. Eric Forsman, the doctoral candidate in biology whose three studies were the *only* studies cited during the listing hearings on the bird, admits to this day that knowledge of the bird is limited at best. According to forest policy analyst Jim Peterson,[2] despite sixteen years of research, no link between old-growth harvesting and declining owl populations has ever been established. Those few logging companies still operational from California to Alaska are required to provide owl habitat and actively manage their lands. They report the highest reproductive rates ever recorded for spotted owls. Two years after the ban, more than eleven thousand northern spotted owls were counted, many of them nesting in second-growth forests and clear-cuts. But the Fish and Wildlife Service would not delist the species. It was fruitless to claim, as many disinterested biologists did, that the northern spotted owl's decline was due to its being preyed upon by the larger barred owl, which had begun moving west a hundred years ago.* Or that federal scientists flatly rejected critiques from biometricians who questioned the statistical validity of evidence on which the listing decision was based. The movement got what it wanted. The largest, most productive, fastest-growing temperate rain forest in the world had been shuttered.

* While Fish and Game have been shooting the barred owl in some states for years, the Obama administration made shooting the barred owl official policy in 2012.

"Culture is far more fragile than nature," said Alston Chase as we bid each other good-bye.

Indeed, the culture of the Washington, Oregon, and Montana forests was pitched almost immediately into trauma. According to timber consultant Paul Ehinger in Oregon, 430 sawmills have closed in the West since 1988, when the war of the woods began. The job losses in the milling and logging industries exceed fifty thousand. And for every forestry job lost, up to five jobs are lost in the businesses and industries that serve the forestry sector. Two hundred fifty thousand family-wage jobs is a mighty blow, and the impacts just keep rolling.[3] Today in Eureka, Montana, there are half as many school-age children entering the system as leaving it. As Bruce Vincent reported, county commissioner Marian Roose told him: "We have a new eight-million-dollar school and we have no idea how we'll pay for it now. Who is going to contribute to our local charities? Who is going to contribute to Little League? Who is going to buy the children's stock at our annual fair 4-H sale? I bet it won't be the attorney for the Ecology Center." For the past decade, the only businesses making money in Ferry County are the U-Haul franchise and storage lockers.

In the nearby Kootenai National Forest, whose 2.5 million acres grows almost 500 million new board feet every year, 300 million board feet dies due to windthrow, insects, and disease. Salvaging and selling that wood could have fed the families of the counties surrounding that forest. "If we don't remove some of this fuel," says Vincent, a third-generation logger, "we are simply stacking three hundred million board feet of firewood in our forest, in our watersheds, around our communities, and around our homes."

In fact, says Holly Fretwell, the shutdown of the forests, the banning of both thinning and the removal of dead trees, even burned trees, has set up a once-in-a-millennium event. As noted in chapter 6, more than 700 million acres of once productive federal and private lands have been set aside under stringent land-use restrictions by federal mandate—almost three times the size of Texas, or 30 percent of the entire nation. Fretwell says that Forest Service experts are convinced that 90 to 200 million of those acres are at risk of cataclysmic fire.[4]

The death of wildlife and endangered species in those fires will far outweigh lives lost to industrialization. Fretwell reports that Victor Kaczynski, a freshwater biologist working on salmon recovery strategies, says, "No single forest practice—not timber harvesting, nor road building—can compare with the damage wildfires are inflicting on fish and fish habitat."[5] Not to mention the dead birds, elk, deer, bears, voles, salamanders, trees and other vegetation, as well as lower summer water volume and increased erosion. The forest can be regenerated. The fish may never return.

Banning thinning has caused 80 percent of the trees in the forests of the Pacific Northwest to become infested with root rot and beetles. The elk and the antelope are gone in many of the forests; the deadfall is too high to climb and the forests too choked to browse.[6] The large predators, wolves and bears reintroduced at the insistence of the movement, are driven by the twenty-foot-high deadfall barriers in the upper forest ranges to forage close to inhabited areas. The resulting loss of sheep and cows can be a blow to a family now merely subsisting. In some rural communities in New Mexico, school bus stops look like cages, and investigations are now confirming wolf kills of humans. The behavior of introduced wolves is changing, and not in a good way. And if the movement, despairing of pure wolf stock in North America, is indeed bringing in Russian cadaver wolves with an average weight of 250 pounds and a taste for human flesh, who among us is going to leave a paved road in the backcountry?

• • •

"I wish," says outgoing county commissioner Joe Bond, "that when these people showed up thirty years ago asking where there was a nice place to eat, we'd pointed them three hundred miles west to Seattle. I say that as a joke, because I enjoy the people who came. But some of those people moved in here and said we don't need the mill anymore. That was when we were running three shifts a day at Boggins Mill and people had good jobs, the kind of jobs that can support a family."

Joe Bond was head saw filer at Boggins Mill for twenty-nine years until Clinton shuttered the western forest. "One guy loses his job, but he has a wife and three kids, so five people move out of the county. Each good forestry job creates between three and five[7] service- or industry-related jobs, so that means if one hundred fifty of us foresters lose their jobs, that means six hundred jobs lost. If six hundred family-wage jobs are lost, then you lose three thousand people.[8] One time our school had six hundred kids; now it's three hundred twenty-five kids. Our hospital is in trouble because the older people who move here to retire, when they go to a doctor, they go to Spokane, so it's just the younger families who use it. Even people my age have left."

Bond's voice is low and gravelly, and I'm not sure—it's late at night, the snow is deep on the ground, getting around was hard today, and we are all tired—but at times I think he is trying to suppress tears. His house is comfortable and well appointed; it gleams through the dark. His wife, Nicole, who runs a beauty salon out of the house, sits with us. They keep a few horses, pets now, since their children had to move to find work. "We watched them take the mill apart," he continues. "It was heartbreaking for our community. After the mill went away, the railroad was bought by a company that bought it for two million and scrapped it out for four million, and rail-banked it with the county. From cutting a hundred million board feet a year, we went to one million."

"What happened to the forest?"

There is a long pause. "It's dying."

He clears his throat. "The forests are warehoused. It's dying, overstocked. Have you been over Sherman Pass? It's that thirty-mile-an-hour corner on the way to Colville. The hillsides up there are turning brown. Where the fire was in '88—that was just a big waste. They wouldn't let the sawmill take the timber, not even three or four years later. A big log-home company from Montana came over, and they wouldn't even let them have the logs for log homes. I talked to a lot of old-timers, and they think that if they had been allowed to go in there with saws and cut a lot of that back down, they would have driven the seeds into the ground and it would have come back a whole lot faster.

"Now we have overgrowth and root rot. I know Okanogan County has moth in their timber; they were trying to spray it, but activists launched a suit to stop it. We know what will happen. The forests in northern B.C.[9] are gone. At one time, we were buying that beetle-kill timber. Now there's no timber coming down from Canada."

His wife chips in. "They want it to be wilderness."

"How likely is that here?" I ask. Wilderness designations are widely feared in working country.

Big, big sigh from Joe. "They keep at it all the time—there was a big turnover in D.C., and wilderness is designated by Congress. It didn't get designated right now, so they have a lot of other fish to fry there."

"Tell her about the dams and what they're doing about that."

"These are small dams on cricks. They're removing them for the purpose of man never ever leaving his footprint on the national forest, easier for them to have their wilderness. Under the 1964 Wilderness Act, it says 'unimpeded by man.' If you're going to treat something as wilderness and there's a crick out there with an eight-foot-wide dam, well it doesn't qualify as a wilderness. So they're ripping them out. Everything dies then. It goes to desert and the water just vanishes.

"They promised that tourism and retirees would make up the difference. As county commissioner, I see all the stats. There's no economic benefit from retirees. Most people who retire here stay seven years and leave. They hit seventy-two or -three, they have to drive to Spokane to see a specialist, they put their houses on the market and leave. There's no tourism. A hundred and fifty people lost their jobs from the mill alone. Those were family-wage jobs that had medical insurance. To create an equivalent one hundred fifty jobs in tourism, you'd have to build a Disneyland."

I drive up to Sharon Shumate's house outside of Republic, Washington, walk in a wide circle around the giant Great Pyrenees wolf-and-bear-killing dog, who is far too stern to submit to my wiles, and settle at the linoleum table. I spend at least half the interview marveling at this woman in her seventies, a sheep farmer, living in an ancient bungalow with a peeling floor and a kitchen so thoroughly unrenovated that I feel shot back in time.

I wake up from my trance when I realize not only that she is schooling me in the history of the movement, but that she is doing it in a manner so clear, so stripped of commentary or emotion, that it is the most succinct I've either heard or read.

"Have you seen the map of what they want?" I nod distractedly, as she presents it.[10] This map, called the Wildlands Project map, was developed from a close reading of the documents detailing the UN's Convention on Biological Diversity, and it rocketed through rural America like a dose of salts.

"That's what they want," she snaps. Shumate has prepared files for me—homework, as it were—detailing the history of the fight in Ferry County. "We fall first because we're the poorest. If we fall, they can take out the rest easily, and not just the rest of Washington state. They'll take out every rural county. They test strategy and tactics here.[11] What they've done is launch multiple attacks on every level, hoping to overwhelm us. It all comes from the IUCN. That's where we've sourced many of their plans."

"Like what?" It seems as if I'm always dragging people off the precipice of paranoia.

"Their priority habitat and species list, for example. They want us to list twenty-nine 'species of interest' here, and our planning commission refused. Some of those species have never existed here, according to their own research. But they think that the species would thrive here, so they want to introduce them—like, for instance, the Oregon spotted frog. Why? And where did this list come from? Fish and Wildlife does not claim to have written it; the Washington State Department of Ecology [DEC] does not claim it. It didn't come from the ESA [Endangered Species Act] bureaucracy either. We finally found it at the IUCN. Just like the biodiversity map. Even if they didn't get that treaty ratified in Congress in 1988,[12] they're implementing it piecemeal.

"So after the Biodiversity Treaty failed, they instituted ICBEMP [Interior Columbia Basin Ecosystem Management Project, pronounced ICE-bump].[13] That's when I got involved, because in 1992, I was head of the Sheep Farmers Association. We defeated ICBEMP, but they put in place the Washington State Growth Management plan right away."

"How do they manage growth if there is no growth?" I ask.

She looks at me with beady eyes for a moment, then throws back her head and laughs. "It's growth for them; they get paid to plan! They submit these templates as models, and that's what gets written as ordinances. They don't acknowledge their linkages and associations.

"The Growth Management Act required each county to develop a comprehensive plan to develop regulations for their critical areas. There are five critical areas defined: aquifer, aquifer recharge areas, geologically unstable areas, floodplains, and fish and wildlife protection areas. These last are the source of greatest conflict.

"At its inception, the Growth Management Act was fairly broad-brush, not overtly threatening to everybody. They offered grants and funding to counties that were doing planning under the act, and they encouraged each county to have a planning department in their county government—provided grants for those county planners; it was how they got into it. Since then, each session of the legislature manages to add a little more to it, broaden the requirements and more teeth to go along with it.

"Now we have forty-eight separate documents for management and conservation programs, status reports, recovery plans, and wildlife research, for which Ferry County residents are supposed to supply habitat designation and document best-available science. Further, nine state agencies are in perpetual motion, buying land to conserve."

Shumate attended agricultural college, then moved to Ferry County in 1988 to ranch sheep. At the time, there were sheep-grazing allotments in the Colville National Forest. But almost as soon as she got set up, the Forest Service began pulling existing leases and refusing new ones.

"It was then we started having our forest fires. Sheep and cattle grazing together keep down all the noxious weeds and brush; they do a fantastic job. Now graze has to be kept four inches high, or animals have to be taken off. Junk science, we call it."

When Sharon Shumate says "junk science," she is referring to the results, because the results are junk. Washington state calls it best-available science, or BAS. There are a half-dozen best-available-management strategies, all of which come from the Codex Alimentarius, out of the World Health Organization and the Food and Agriculture Organization of the United Nations. Codex Alimentarius claims to set international standards for "good safe food"; what it does is overturn generations of local rural knowledge, its eventual aim, claim rural people, to police all food production and force it to conform to international standards.

"So-called best-available science tends to be kind of consensus science—some people at an agency, say Fish and Wildlife, come up with a theory, think it's maybe kinda true, and promote that as their science," says Shumate. "Take the battle we have been fighting for years now: one or two hundred feet buffer width on streams. They want to keep water temperatures at a certain level; they want shade for the water; they want plant life on the east side of the shores; and they want large woody debris to be able to fall into the creeks. Those conditions are promoted as 'best-available science.'

"It means essentially, 'Let's see what everyone else is doing and we'll put together a synthesis and publish that as best-available science.' Commerce doesn't distinguish; it treats that document as the absolute gospel truth, despite the fact that the authors say right in it that it's not meant to apply specifically. After Commerce has put its imprimatur on the 'science,' the rest of the agencies treat it as the scriptures of God. If they get it put through here, then they can tout it as being in use in other places."

Ferry County has thrown up some of the most effective fighters for rural people in the country. In 2000, Lenape Indian Diana White Horse Capp published a book called *Brother against Brother: America's New War over Land Rights*. In it, Capp—who testified before Congress in 2000 and received multiple death threats for her efforts—says that environmental activists stirred up the tribe on the Colville reservation and paid them to support the movement's objectives in Ferry County. Indians on the Colville reservation have intermarried with whites for more than a hundred years, which meant families were being torn apart.[14]

Ten years on, three legislators fight every step of the way. Shelly Short in Washington state is advancing a bill that would require that each piece of science-based regulation affecting rural Washington state receive independent peer review. Defeated in April 2011, it will almost certainly pass in the next session, the first such legislation in North America. Representative Joel Kretz forced the legislature in Washington to reject the Yellowstone-to-Yukon wildlife corridor. In D.C., Cathy McMorris Rodgers fights the good fight against a tidal wave of money and lobbyists.

I visit Kretz and his wife, Sarah, in their house in Ferry County. It is a typical rancher's house—timber frame, worn, comfortable, with a three-mile driveway. Kretz and his wife breed mustangs, do a little timber harvesting, and run a score of mother cows.

"Yeah, I remember Mitch Freedman from Earth First throwing horseshit down the ventilators at the mills. I got involved a few years ago, when we were doing short loads, logging with horses, just light thinning, stand-improvement stuff. Our culverts were working fine. Then some guy from the Department of Natural Resources comes up and says 'as part of your privilege of being able to harvest your timber, we want you to put in a twelve-thousand-dollar arc-style bottomless culvert to save salmon.'

"Now, four things have to happen before I save salmon in Bodie Creek. They gotta jump the Grand Coulee Dam, they gotta jump the Chief Joe Dam, they have to jump the sixty-foot waterfall on the Kettle River, and then they have to cross a quarter-mile of rocky ground to get to this culvert so I can save it. Plus, it's a seasonal runoff ditch; it only holds water for six weeks a year.

"So," he continues, "after we called them on that, it became bull trout, which they introduced into the creeks and called native and endangered. All of us on the creek got letters saying you have thirty days to comply or daily fines will be levied. So I called Fish and Wildlife, and this little prick biologist came out, stuck his finger in my face and told me, 'You've made a lot of money off this land. It's time you gave back to the environment.'

" 'Well, comrade, it's not working for me,' I replied.

" 'I'm just doing my job,' said he.

" 'That was a really popular response in Nuremberg,' I told him. 'Most of those people got stood up against the wall.'

"Now he's the chair of the local land trust.

"I'm providing lots of public benefit here. There are lots of species and wildlife on my land that are doing really well. Why doesn't the public owe me? As chair of the land trust, he'll acquire land and let it sit. Within months, it will be a noxious weed patch or, if a forest, there will be deadfall, pests, brush, mistletoe. There's no grass in the forest anymore; shade shuts it out. The forest is overstocked; we have hundred-year-old trees that are starving, dry. Which means when the forest burns, it burns hot.

"There's a biologist under every rock out here busily protecting wildlife from the peasantry: the evil, unwashed Neanderthals."

Sarah interjects, "We have a photo somewhere of a sign that says 'Fish and Wildlife Refuge.' It's surrounded by acres of knapweed. No wildlife can survive in that. It's barren ground until it's taken care of."

Sarah's favorite saddle horse was shot shortly after this incident. "Both gates were locked, so someone hiked in three miles, at night, and shot her between the eyes. They knew which horse because she was on the website.

"Then they banned hunting with hounds. Hounds are the only way to deal with mountain lions. Otherwise you lose one colt or calf or pet after another—I've lost twenty colts, found them in pieces. The biggest drop is the conception rates, a ninety-two or ninety-four percent conception rate drops to fifty percent when there's lion around.

"Right now, if there was a lion on that deck, I couldn't shoot it. So if a big cat is on my horse, I'm supposed to call Fish and Game. Which is a good hour's drive from here. They issue pamphlets: *Living with Cougar.* Apparently you're supposed to put your hands up in the air and look big. Are you kidding me? If there was a mountain lion or bear in Seattle, they would call out SWAT teams, the National Guard, and helicopters. I'm all for spreading the love; let's introduce wolf and bear into the Seattle area. They've introduced two wolf packs into next-door Stevens County. They want grizzly next. Maybe a few grizzly wandered through here a hundred years ago, but they want them introduced anyway.

"If they treated a racial minority the way they treat and talk about people out here and me in the legislature, they'd be in trouble. But it's all right to talk trash about the rural peasantry."

"I'm not sure why they let us live," says Sarah.

"So far," replies Kretz. The room collapses in a fit of gallows humor.

An Abundance of Frogs, Snakes, and Water

In the meantime, I wait. And wait. While money trickles away. To the banks, to consultants, to the five[1] lawyers I have to hire, in the incessant demand for fees from the bureaucrats, surveyors, and engineers.

One day, when Kathy visits to talk about the salmon creeks, I mention that I've seen several small, sweet, blue-listed, red-legged frogs in the past weeks, and those blue-listed, sharp-tailed snakes? There could be a few less of them.

"Oh no you didn't!" she shouts at me as we trudge up the trail.

It's a joke, I think, so I laugh and persist. "Oh yes I did."

She turns around and for once is stern and direct. "No, you didn't. And don't tell me—or anyone like me—again."

I get it. My restrictions would proliferate. But if I, a rank novice, nearsighted and usually lost in the clouds, can find endangered species so easily, how much of this species death is real?

Every morning, after staring at my list and calculating whom I can call without being moved to the nuisance list and further punished, I slide down the hill with the dogs to the sixteen acres I will keep. Work on my lowly patch is progressing. Today the salmon pond is being dug, and the water witch is coming.

I hear the racket halfway down the hill. Close to the road, an antique orange ten-ton excavator is slicing long curls of black dirt, thicket, and brush out of the future salmon pond. Mint fills the air; it was one-half of the pond-site cover, and it had gone to flower. Over it all, the roar of compressors and the crashing as small trees break and fall, hovers the essence of what is dying most and fastest.

I cut broom at the back of the meadow until Albert Kaye arrives.

One possible bankrupting expense is the new well that must be dug on the new property. The islands are notoriously dry, or at least so goes the propaganda. I have called around and found estimates that range from $4,000 to $15,000 and have chosen, of course, the former, the longest-established driller on the island. Like his father and grandfather before him, he also witches water.

Albert climbs down from his big truck, looking for a moment like any other tradesman in middle age, thickened around the waist, with roughened skin and well-worn clothes. Then he turns and beams radiance. His face is soft and round, seemingly shining with innocence, with big blue damp eyes that are very kind.

He traverses the meadow while I follow him. "I've dug almost all the wells up and down this road," he says over his shoulder, while he saws a forked stick from a tree with something so small, it looks like a penknife. "This place is crisscrossed with underground rivers."

Then he starts to pace the land, holding out the stick. From time to time, it flips down and points at the ground. Albert flags the spot. I think he's faking until he lets me hold one tine of the fork while he holds the other. The magnetism nearly rips the stick out of my hand. When I let go, my palm is scraped. I stare at him in wonder.

"On this property, you can choose where you want the well, and I can find water," he exults.

I choose the cheapest, easiest location, near the road. The next day he brings in the well platform and two hundred feet down finds an artesian gusher that spurts water in a glorious unending stream. When Kathy sees it and the water filling up the pond, I get to watch her dancing again. "The water witches of Salt Spring strike again!" she cries.

Here's my question. If there is supposed to be a drastic shortage of water on the northern Pacific coast (on the face of it, risible) and two threatened or endangered species (they move from red to blue to red), how on God's green earth do I have an abundance of all three?

What the hell is going on?

CHAPTER TEN

Where Are All the Corpses?

The rationale underpinning this subterfuge lies in a form of science established, more or less, in 1978, at the University of California–San Diego. Conservation biology marks a sharp veering away from the scientific method we all learned in middle school. The scientific method is the reason bridges don't collapse when you drive over them and that 99.999 percent of the time, your food is not dangerous to eat. A hypothesis is advanced, and the relevant scientific community bends toward testing it. The scientist has to go back, revise some of his thinking, and advance his theory again. And when every shred of falsification or doubt has been rooted out, the theory is deemed to have utility.

Conservation biology starts with the assumptions that humans are wreaking havoc on the natural world and that resources are finite and decreasing. These are plausible assumptions, but assumptions nonetheless. History and hard evidence demonstrate that bounty, life expectancy, health, wealth, clean water, and air are all *increasing*, at least for those of us who live in countries under the system of democratic capitalism. Oil finds, for example, in 2010 and 2011 surpassed even optimistic predictions by an order of magnitude, at which point the peak oil argument went all quiet. Next comes the assertion that a third of all animal and plant species are threatened, largely because of "development" and corporate greed.[1] The job of the conservation biologist is no longer investigation but educating the public on biodiversity loss and species extinction. Conservation biologists claim that their profession holds higher values and furthermore, those values are universal, so they must be treated as normative or established fact.[2]

Until the establishment of conservation biology, biologists had increased the bounty of agriculture and our understanding of how health is produced; they were regarded, as economist Julian Simon famously said, as the profession that had contributed most to human well-being. Conservation biologists promptly hijacked that reputation.

Old-school biology told us that nature is a complex system that we are only beginning to understand. Conservation science said no, we do understand: nature is made up of definable ecosystems, which must be in perfect balance, each member not only in effulgent health but present in the correct proportion.

To old-school biologists, that was observably wrong. You can take the buttercup and deer off the meadow, and the meadow will not only survive but thrive. By 1995, the phrase "lowly patch" was gaining currency. Lowly patches could be seen bleeding into each other everywhere you looked. A meadow can change from desert to swamp within a hundred yards, depending on the hydrogeology and earth substrata. Further, not only did nature change, but more often than not, the change was rapid and catastrophic. As Alston Chase points out in *In a Dark Wood*:

> *A host of studies revealed that creatures of every kind, from insects to elephants, undergo extreme and often destructive fluctuations. Records of German coniferous forests show that the numbers of lasiocampid moths sometime multiply by a factor of over ten thousand in two years. Chinch bug populations along the Mississippi rose and fell erratically and dramatically between 1823 and 1940.*

> *Moose and wolf populations also were found to vary wildly over time. . . . Trappers' records, Botkin reports, establish that Canadian lynx populations oscillated between eight hundred and eighty thousand over a 240-year period ending in the 1940s.*[3]

In fact, change in nature was the *only* constant, which meant caretaking was possible. Do you want wildlife? Then this is what you do. Do you want a stunning landscape? Well, these are the actions you take. Do you want a working forest? A cattle range? Open space?

But if man destroying the natural world is a first principle, then what you look for is destruction, not evidence of health. And where you are going to find destruction is in the footprint of man. This is circular reasoning that we have all apparently adopted as established fact.

The solution? Remove man. Not necessarily you from where you are, but other people from where they are.

And if you are working from a construct that does not exist in nature, but only in computer models—that is, the ecosystem—then models become everything. With the Nature Conservancy's NatureServe.org, the U.S. Geological Survey's ten-dimensional nature mapping, and the hybrids used by the International Union for Conservation of Nature, saving the biosphere turns out to be desktop work! Hello, how convenient is that? No need to deploy expensive armies to fact-check *assumptions.* All you have to do is plug the "findings"[4] of the Intergovernmental Panel on Climate Change—also largely based on modeling—into the desktop, and presto, 33 percent of species are definitely about to go extinct.

Thirty years after the founding of conservation biology, with trillions spent worldwide to preserve species and biodiversity and with hundreds of millions of lives foreshortened and curtailed to preserve the natural world, things don't look so good. In fact, they are worse. Despite setting aside vast amounts of land all over the world—an area greater than the size of the continent of Africa—and despite restricting most development to only 3 percent of the American continent and locking down in heavy regulation 90 percent of the rest of the landmass, species are still dying. Hand over fist. Faster than we can count.

The results of this "thinking," however, are actual rather than virtual. In just thirty-five years, conservation biology has created one disaster after another, in something that observers are now calling an error cascade. Tens of millions have been removed from their beloved lands. Immensely valuable natural resources have been declared off-limits to the most desperate in the developing world. In America, cattle have been removed from millions of acres of range, 925 dams have been removed from the once magnificent waterworks of the United States, 90 percent of the Western forest is off-limits, and the countryside is emptier than it has been since the beginning of the twentieth century. Range, forest, and farm are dying; water systems have been destroyed. Conservation biology has created desert and triggered the dying of entire cultures.

The aim of old-school biology was to create bounty, using evidence-based management. Increasing bounty by cutting or planting trees, removing invasive species, planting crops, bringing in or removing livestock, and increasing water flow were fruitful activities. The relatively new science of ecology—decidedly *not* synonymous with ecosystem theory—depended upon higher math rather than professed virtue and sentiment. Ecology was related to physiology, ethology, and genetics; it worked solely with the lowly patch and, together with the tools of the digital revolution, it promised increase of Edenic proportions. That didn't happen, but if we can pry the cold dead hand of the conservation biologist from our natural world, it still could.

A few brave souls stick up their heads now and then. In *Science and Public Policy*, Professor Aynsley Kellow, head of the school of government at the University of Tasmania, describes a paper published in *Nature* in January 2004 that "warned of the loss of thousands of species with a relatively small warming over the next century. But just how virtual was this science is apparent when we consider that the estimates of species loss depended upon a mathematical model linking species and area."

In a follow-up piece called "However Virtuous, Virtual Science Is No Substitute for the Real Thing," Michael Duffy pointed out that "Kellow notes that a similar warming over the previous century had not left anything like the trail of species devastation being proposed in the paper, yet this observational data was considered irrelevant compared with the virtual world of the models."

In point of fact, since the white man arrived in the Americas, there have been no *forest* bird or mammal extinctions from any cause.[5] Not even one. The Pacific Research Foundation, which is the least hysterical environmental auditor, finds, sourcing from the IUCN Red List and the occasionally hysterical Heinz Center's State of the Nation's Ecosystems project, that only 2.7 percent of species have gone extinct since the last ice age. In 2001, in a *Nature* essay called "Has the Earth's Sixth Mass Extinction Already Arrived?" fourteen scientists agreed that only a tiny fraction (2.7 percent) of species had been formally evaluated for extinction by the IUCN.

What we're frightened of is desktop extinctions, not real extinctions. And in fact, so much time has been spent on virtual science, we don't even know how many species there are in the world: estimates range from 1.5 million to 40 million. New species are being discovered every year by the thousands.

In the late 2000s, several respected writers went haywire and predicted the collapse of the biosphere. Jared Diamond in *Col-*

lapse and Ronald Wright in *A Short History of Progress,* based on his five 2004 Massey Lectures, used Australia as an example of Western agricultural science laying waste to natural biological systems. These attacks inspired several cross but nonetheless detailed refutations,[6] pointing out not only that Australia was doing just fine, thank you, but that the writers had missed the critical point that island extinctions are always greater than the norm by several orders of magnitude, and that one cannot project island ecology onto continents without producing grievous error.

According to Norman MacLeod, executive director at the recently founded Environmental Sciences Independent Peer Review Institute[7] in Washington, D.C., the Fish and Wildlife Service likes to manage everything by subspecies and distinct population segments. But there are a considerable number of biologists who don't agree that you should manage anything by subspecies or even that there is validity to subspecies, much less distinct population segments. It is likely that the very slight genetic differences in fish from one creek to another—which is what is meant by "distinct population segments"—are unnecessary for the survival of the species as a whole. But Fish and Wildlife manages those creeks as if each were absolutely critical.

"Distinct population segments are big in the salmon-saving world," says MacLeod. "So today each salmon stream has its DPS; so do bull trout streams. It's a nice argument, but as to whether it holds water, it probably doesn't.

"Here's the point. Is the goal to protect habitat or prevent human activity on the landscape? If it's to prevent human activity, the threatened species gets it in the shorts every single time."

We live in a world where environmental policy is based on the following logic: despite the fact that we have no idea how many species there are by at least one order of magnitude, and despite the fact that only 2.7 percent of known species have gone extinct since the last ice age, we nonetheless believe that one-third of the species on the planet are about to go extinct.

How is such a thing possible?

Answer: Harvard professor Edward O. Wilson. His theory of species-area relationships still dominates; he is the colossus astride extinctions. Hold on for some real sophistry.

Willis Eschenbach writes:

> *In their seminal work, "The Theory of Island Biogeography," MacArthur and Wilson further explored the species-area relationship [Robert H. MacArthur and Edward O. Wilson,* The Theory of Island Biogeography *(Princeton, NJ: Princeton University Press, 1963)]. This relationship, first stated mathematically by Arrhenius in 1920, relates the number of species found to the area surveyed as a power law of the form* $S = C * a ^ z$*, where "S" is species count, "C" is a constant, "a" is habitat area, and "z" is the power variable (typically .15 to .3 for forests). In other words, the number of species found in a given area is seen to increase as some power of the area examined.*
>
> *By surveys both on and off islands, this relationship has been generally verified. It also passes the reasonability test—for example, we would expect to find more species in a state than we find in any one county in that state.*
>
> *Does this species-area relationship work in reverse? That is to say, if the area of a forest is reduced, does the number of species in the forest decrease as well? And in particular, does this predicted reduction in species represent species actually going extinct? One of the authors of "Island Biogeography" thinks so.*
>
> *In 1992, E. O. Wilson wrote that because of the 1% annual area loss of forest habitat worldwide, using what he called "maximally optimistic" species/area calculations, "The number of species doomed [to extinction] each year is 27,000. Each day*

it is 74, and each hour 3." [Edward O. Wilson, The Diversity of Life *(Cambridge, MA: Harvard University Press, 1992)].*[8]

In 2010, Willis Eschenbach fact-checked this and other worrying assertions and published his findings in an essay called "Where Are All the Corpses?" that burned through the Internet like Agent Orange. Eschenbach writes, "If we have lost 27,000 species per year since 1992, that's over 300,000 species gone extinct. In addition, Wilson said that this rate of forest loss had been going on since 1980, so that gives us a claim of over well over half a million species lost forever in 24 years, a very large number."

In fact, however, of the 4,428 mammal species (IUCN Red List 2004) living in Asia, Europe, Africa, North America, South America, and Antarctica, according to the IUCN, only three have gone extinct in the last five hundred years: the bluebuck antelope in South Africa, the Algerian gazelle in Algeria, and the Omilteme cottontail rabbit in Mexico. We see the same pattern with birds as with mammals. Of the 128 extinct bird species, 122 were island extinctions. Of the 8,971 known continental bird species (Red List 2004), 6 have gone extinct worldwide, with only 2, the passenger pigeon and Carolina parakeet, in the U.S.*

As for forest extinctions created by habitat loss, Eschenbach found *none.* Furthermore, he wrote, the most recent total bird and mammal extinction rate in *all* parts of the world, including islands and continents, stands at 0.2 extinctions per year. This is down from a peak of about 1.6 extinctions per year a century ago.

Wilson, with his species-area relationship, which is in use now in every conservation data center in the world and is *the*

* If the range runs from 1.5 million to 40 million, and thousands of new species are found every year, we don't even know the denominator of the fraction. So how on earth do we predict an extinction rate? Stephen F. Hayward of the Pacific Research Institute puts the variance in the number of species at two orders of magnitude. Source: *2011 Almanac of Environmental Trends,* p. 237.

dominant equation used in making land-use decisions outside the cities, claims extinction rates two hundred times higher than the data shows.

So I can have someone for dinner who is funny, warm, smart as a whip, with a serious doctorate from a serious university, and after a few glasses of wine, she's tipping my Charles Rennie Mackintosh chair back onto two legs and insisting at the top of her voice that ten thousand species a year are going extinct.

Such is the corrupt, lazy, destructive, observably *wrong* thinking that underlies the Endangered Species Act, a law that routinely overturns the Constitution of the United States and is the weapon of fire used by the movement to destroy rural America. Rural people say that because of the hysteria around the species loss question, we do not know what we are losing and, therefore, cannot address the loss using adaptive management.

The cost of the Endangered Species Act to the general public ranges up into the region of hundreds of billions of dollars, and that is not counting the certain trillion lost in opportunity cost.[9] As of 2010, there were 1,967 species listed as threatened or endangered, 98 proposed, and 249 candidates. According to the U.S. Fish and Wildlife Service, which administers the ESA, the cost of listing a species is $85,000 and the average cost of designating critical habitat is $515,000. Overall, listing and habitat have cost about a billion dollars, which would sound reasonable if the listings were not based on agenda-driven science.

Never mind, we've just created a pot of honey so large that entire legal careers have been built around the contradictions inherent in the terrible science and the untested assumptions that now determine the use of every resource we have. Litigating the Endangered Species Act is an enormously profitable activity and represents a large part of many ENGOs' annual revenues. When lawmakers passed the Endangered Species Act in 1973, they filled it with deadlines to force bureaucrats to make timely decisions. Karen Budd-Falen, a lawyer and fifth-generation rancher in Wyoming, writes, "Earthjustice Legal Foundation (a public interest, nonprofit legal foundation) representing Defenders of Wildlife, Sierra Club, The Wilderness Society and Vermont Natural Resources Council has filed an application for attorneys fees in a single case that took 1 year and 3 months to complete for a total of $279,711.40. *For that same suit,* Western Environmental Law Center (also public interest nonprofit legal foundation) representing Citizens for Better Forestry, Wild West Institute, Gifford Pinchot Task Force, Idaho Sporting Congress, Friends of the Clearwater, Utah Environmental Congress, Cascadia Wildlands Project, Wild South, Klamath Siskiyou Wildlands Center, the Lands Council, Forest Service Employees for Environmental Ethics, Wild Oregon, and Wild Earth Guardians filed an application for attorneys fees for $199,830.65. Thus in TOTAL environmental plaintiffs are requesting $479,242.05 for a single lawsuit lasting 15 months. . . . [B]etween 2000 and 2009, only 8 environmental groups filed 1,596 lawsuits against the federal government."[10]

Any group or individual can petition to list a species, and the Fish and Wildlife Service must reply within ninety days. In the late 1990s, the Southwest Center for Biological Diversity sued Fish and Wildlife for missing the deadline on forty-four rare California plants and won on all forty-four counts.[11] In every case, the petitioner's legal fees are paid by the service, or rather the taxpayer. Hundreds of millions of dollars and thousands of hours of bureaucrat and court time are spent every year to satisfy this one requirement.

Budd-Falen wondered just how much money the ESA cash cow has meant to environmental groups and bent herself to some forensic accounting. She found that "since 1995, there has been no accounting of this money, money that by the smallest of estimates is in the hundreds of millions of dollars," paid to groups like the Sierra Club, the Center for Biological Diversity, Western Watersheds, the Wilderness Society, and Defenders of Wildlife. In fact, says Budd-Falen, there is endemic cooperation between activist litigators and agency bureaucrats.[12] "My friends sue me and I can force new regulation," she says. "That's how it goes."

Budd-Falen then dug into court documents, first physically visiting and reviewing case files in various courts, then examining the dockets and pleadings for every environmental case using WESTLAW and PACER, data services that respectively archive published and some nonpublished federal decisions, and dockets and filings for every environmental case in federal court in recent history. As far as her initial estimates of the cost of this litigation, she was off by one order of magnitude; in fact, the number is way up into the billions. Budd-Falen first counted the amounts awarded ENGOs under the Judgment Fund, a line-item appropriation used for Endangered Species and Clean Water act cases. In attorney's fees alone, the Judgment Fund paid over $4.7 *billion* from 2003 to 2007 to various environmental groups. On a smaller scale, between 2003 and 2005, the six regions of the Forest Service alone paid out $1.6 million in legal fees to environmental groups under the Equal Access to Justice Act. Keep in mind that these were exclusively awards given to groups because the Forest Service missed a deadline or filed a brief with a minor omission. In many of these cases, the environmental organizations were paid legal fees just because they filed a complaint.

"Activists love to cite the fact that it was a Republican Congress that struck a line through the need for accounting in 1995, but I think they didn't know what they were doing." Budd-Falen lobbied to reinstate the accounting, and her bill survived the committee process and was heading to the House at the end of 2011. "There is no oversight of this money, and that money is coming out of budgets that should be funding maintenance and protection of public lands, national forests, and national parks," she says. "The billions received by nonprofit tax-exempt environmental groups is not spent protecting wildlife, land, plants, animals, or people, but is being used to fund more lawsuits." The process has grown so perverse that even an activist with Defenders of Wildlife worried that not only do the suits have nothing to do with the species or even its habitat, but they are coordinated to overwhelm the system and force a significant reshaping of society.[13]

Tom Knudson was the first mainstream reporter to peel back the pleas of poverty and smarmy "of the people," "grass roots" posturing of the movement. Knudson demonstrated that the movement is a monolith of money and power working with the most sophisticated financial instruments and the biggest marketing firms and funded by the largest private fortunes the world has ever known.

Knudson's most valuable contribution was his groundbreaking reporting on the legal looting of the U.S. Treasury by environmental groups. He covers the issue of missed deadlines and egregious lawsuits but goes deeper into the corruption by detailing the critical-habitat racket, which has locked up tens of millions of acres. Frogs—red-legged ones, like mine—got one-twentieth of California, or 5.4 million acres, in the late 1990s, and today in every rural area in the world, *all* land, whether around water systems or not, is either already locked down in a maze of regulation or under imminent threat of critical-habitat designation.

Even if the critter has not been found in an area, if the movement thinks it should be, it will be, as for instance, the frog from southern Oregon that Fish and Game wants to introduce into the magical forests of northeast Washington. Habitat will then be designated, whether on private land or not. The idea is that the newt or frog could have been in residence once or, if introduced, could possibly recolonize the area or even merely pass through on its way to somewhere else, so land and water must be frozen and all commercial activity must stop. Around Tucson, 790,000 acres of critical habitat for the cactus ferruginous pygmy owl forced smart growth, freezing exurban land forever. The desert tortoise has hobbled all development in the Arizona and California deserts, and investor Robert F. Kennedy Jr.'s BrightSource Energy is spending, as of April 2012, $40 million to move the desert tortoise from its solar farm. It's all modeled on the ill-founded listing of the spotted owl, which curtailed 90 percent of the economy of forested communities in the West, acting as trigger to a negative cascade of poverty and drug addiction, family breakup, and forest destruction.

The movement has spent the last ten years trying to list the prairie dog, which is so common, it's like the rabbit of the plains. Prairie dogs certainly breed like rabbits. Ranchers think of them as massively destructive rats, because they can eat and dig a field or range into desert within two months unless they are controlled. So every year, prairie dogs are poisoned to limit their damage so crops can be grown. But the movement has been persuading vulnerable types to release the endangered black-footed ferret on their farms. That means that every time farmers or ranchers poison prairie dogs, they first must round up every black-footed ferret in the area, or else their fines will be breathtaking. It is very difficult to find the ferret; it hunts at night, so farm kids thrash through the scrub for weeks to capture them alive.

Pleistocene introductions are gathering currency, so the Bolson tortoise, an animal found in only a small part of Mexico, is being introduced in the Sonora Desert, and habitat is being designated for it. The Center for Biological Diversity is agitating for the introduction of the jaguar in New Mexico's boot heel, where jaguar experts assert the cat has never lived.

The two worst offenders, though there are many candidates, are the Center for Biological Diversity and the Western Watersheds Project. The Center for Biological Diversity claims to have caused the listing of hundreds of species and has spent the last few years developing plans to list another seven hundred. Between 2000 and 2009, the CBD filed 409 lawsuits and 165 appeals in federal courts. The center has been the recipient of tens of millions in legal fees from the government and was at one time so frenetic that it filed a new lawsuit every thirty-two days. Western Watersheds, an ENGO started and run by Jon Marvel, a society architect and proud Vietnam draft dodger who lives in Hailey, Idaho, runs a close second. Marvel filed ninety-one lawsuits and thirty-one appeals between 2000 and 2009. Marvel claims that he has "saved" millions of acres of ranch land from the predations of cattle and that his priority is to get cattle and sheep off the American range.

Karen Budd-Falen cites the notorious statement of Kieran Suckling of the Center for Biological Diversity in 2009 in order to further illuminate the destructive intent behind lawsuits launched under the ESA:

> *New injunctions, new species listings and new bad press take a terrible toll on agency morale. When we stop the same timber sale three or four times running, the timber planners want to tear their hair out. They feel like their careers are being mocked and destroyed—and they are.*
>
> *So they become much more willing to play by our rules and at least get something done. Psychological warfare is a very underappreciated aspect of environmental campaigning.*[14]

In Lander, Wyoming, a town of 7,400 on the southeast slopes of Yellowstone National Park, I happen upon a meeting of the county commissioners. Fremont County, the seat of which is Lander, is home to 40,000 people, who ranch, farm, mine, and work on the natural-gas rigs. The county is big, 9,300 square miles.

There aren't that many people in attendance—a few ranchers, a man who turns out to be from Encana, the Canadian energy company that is deep into natural-gas extraction, a congressman, a state senator, and me. Around the table set in front of the commissioners sit bureaucrats from the Forest Service, Wyoming Fish and Game, the Fish and Wildlife Service, the Bureau of Land Management, and the Wyoming Oil and Gas Conservation Commission. At issue is the threatened federal listing of the sage grouse, a bird once so common you can think of it as the robin of the range. "Wyoming," says one commissioner in protest, "is nothing *but* sagebrush and sage grouse." Nonetheless, the Wilderness Society, by some convolution of the species-area equation, is bent on listing the fowl, whereupon all activity around sage grouse habitat—in other words, all of Wyoming—will be curtailed. Please understand that the sage grouse might be decreasing in numbers—more neutral observers point out that there appear to be fewer; however, that is the case on pristine habitat as well as on rangeland in some sort of use. The common denominator seems to be the introduction of wolves and coyotes,[15] the behavior of which is definitely changing since they discovered how much easier it is to hunt the grouse, farm animals, or the family dog than bring down elk or deer.

The bureaucrats present today think they have found a solution to the problem raised by the Wilderness Society. Fremont County, the Fish and Game bureaucrat informs the county commissioners, has 85 percent of the core habitat of the sage grouse.

He unfolds a map in front of those commissioners, and then, for the audience, uploads the map onto a screen.

Here's the thing, he says. Unless the county radically curtails activity on its range, the Wilderness Society will force the listing of the bird and *all* of Wyoming will be in trouble. Fremont County is being asked to take one for the team.

The expression on the faces of the county commissioners is one of dumbfounded horror, and immediately all the bureaucrats go into their song and dance, as practiced a routine as you would find on Broadway. A bunch of studies and management plans are trotted out and praised to the skies. We did the bighorn sheep plan, says Fish and Game; we discovered the best of the best habitat, and that worked. We are mapping, says the Forest Service, and hey, it looks as if those maps overlap and some of the core sage grouse habitat is bighorn sheep habitat, so score! We've decided that 80 percent of the birds breed in four-mile circles, and all we have to do is buffer those circles! We will do "truthing" with known permits, and there will be grandfathering! We will include local people, in fact; if we're going to get this done, we'll get it done locally. We have sagebrush mapping in progress! We will keep ranching whole! We will work with ranchers; there will be no prohibition on grazing—no, not at all. And so endlessly on.

Finally, one of the ranchers leaning against the wall at the back of the room asks in his outdoor voice, which manages to freeze every functionary at the table, "What if you're in the core area? Can we develop water?"

A pause. "No."

"Can we drill?"

"No."

"What about opportunity? What about wind, gold, uranium, oil, and gas?"

The relevant bureaucrats are polled.

"No."

"No."

"No."

"In core areas, there will be greater standards, but we will have flexibility."

"But the core areas you identified have a lot of potential in oil and gas, and you have already pitched our grazing base into decline. Restrictions only ever increase," says the rancher.

"They are threatening a federal listing; we were in a very vulnerable condition," says U.S. Fish and Wildlife.

"When you found eighty-five percent of the core habitat in Fremont County, that would have been a good time to get us involved," says Pat Hickerson, one of the commissioners. "Not after you made a deal with the Wilderness Society."

It is clear who the boss is in this room and in Fremont County. It is the Wilderness Society and its proxy, the Endangered Species Act. With the stroke of a pen, the Wilderness Society has diminished the livelihood and future of forty thousand people, apparently without thought or care, for a fowl once so common, it might as well have been a Manhattan cockroach.

Faced with the resistance of county commissioners, ranchers, and politicians, Fish and Game admits by the end of the meeting, "The more we study the sage grouse, the more we realize that they are thriving, and everywhere, even around tailings ponds. All we're asking is a conservation strategy with assurances."

"That doesn't make me feel real warm," says Pat Hickerson.

"Do we just want to commit suicide now or just let them kill us?" asks his colleague, Commissioner Dennis Christensen.

This time, no one in the room thinks that's funny.

In Washington state, one of the Nature Conservancy's surrogates, the Center for Natural Lands Management (CLM), a California-based nonprofit, along with the Cascadia Prairie Oak Partnership, is heading a drive to preserve diminishing prairie habitat. The habitat is diminishing, says Norm MacLeod, because the Indians used to burn the forests to create prairie. Since the movement stopped logging, which created prairie much as Indian burning did, prairie has been vanishing in the Northwest. But never mind that inconvenient truth. CLM is using the so-called threatened prairie/pocket gopher and its diminishing habitat, *which they diminished by stopping logging,* specifically to stop farming, ranching, and building on any open space, or rather "pre-Columbian prairie," in the region.

"The principle here, using the prairie/pocket gopher example, is that if you focus too much on restoring to pre-Columbian conditions any particular landform, and the species actually thrives far better in a different habitat type than in your focal habitat type, you can actually end up depressing its overall population," says MacLeod.

In fact, despite CLM's assertions, pocket gophers actually do better in what is called early successional habitat postharvest and replanting—commonly known as Christmas tree farms and clear-cuts—than they do in mature prairie habitat, but, says MacLeod, "if you are basing your population numbers by only those gophers found in prairie settings, you do get low numbers, low enough to 'justify' listing. If, on the other hand, you do population counts in neighboring recovering clear-cuts and commercial Christmas tree farms, you come up with an entirely different picture."

If you then list the gopher or any of its subspecies as threatened or endangered based on its presence in the prairie landform you are trying save/preserve/recover, and if that landform is actually used only as a "reserve" habitat by the gopher, you signal to the human community that it's not in one's best economic interests to have gophers around. If you own commercial timberland or make every dime you have from your Christmas tree farm, and if you know that the enforcement for gopher livelihood is focused almost exclusively on the prairie in your region, what are the practical implications?

MacLeod continues, "Your income is derived from your forests or farm. These are not recognized as critical habitat for the gopher. Nonetheless, your property is infested (from your perspective) with thousands of gophers. Recognizing that there is a great variation in ethical conduct in the world, and the landowner's bottom line is at stake, what do you think is a good chance of happenstance for the gophers in the clear-cuts and Christmas tree farms?

"And if the landowner gets caught doing for the gophers, what's the likely response? 'Oh, I'm sorry, I thought I was dealing with a mole problem.' "

At that point, one's mind veers away from the fantasy we are forced to accept as truth. But to refocus, if CLM's goal is to preserve prairie for the sake of the gopher, yet prime gopher habitat is the clear-cut or Christmas tree farm, doesn't that have the potential for a wildly inaccurate population count as the basis for an improper listing? And if the Wilderness Society has the muscle to drastically limit commercial activity on the range because the sage grouse are disappearing due to wolf introduction and forest overgrowth, both of which *the environmental movement created*, we have truly lost any hold on reason.

Then, if the fact that official eyes are cast upon the wrong habitat type as being critical for the gopher's survival, and if thousands upon thousands are getting mole treatment to prevent damage to commercial tree seedlings, don't we have the potential for the gopher to become threatened or endangered in fact, as well as in public perception?

Shoot, shovel, and shut up. Or as Kathy Reimer said to me . . . oh, right, she said the same thing: "No, you didn't. And don't tell me—or anyone like me—again."

On September 29, 2011, the U.S. Fish and Wildlife Service decided to put another five hundred species on the list, including thirty-five different kinds of snails in Nevada's Great Basin, and eighty-two different kinds of crawfish in the Southeast. In Hawaii, ninety-nine new indigenous plants must not be "disturbed." If the "science" on any of these is as sound as that on the sage grouse and prairie gopher and if it's based on the species-area equation, it's all corrupt. It's all nonsense. Who knows what's threatened and what isn't? We know economic activity slams up against a wall and dies. But without accurate science, we could be losing something with actual, rather than imagined, value. We don't know. And we're out of money, since we've killed the tax base, so we can't afford to determine the truth of it. The science is so flawed, it's destructive beyond measure.

We have to throw out the Endangered Species Act and start the science all over again.

Way Out on the Edge of Complexity

In early December, I arrive at the Islands Trust office at 10:00 A.M. Denis Russell meets me in the parking lot. Despite being retired, he still has his engineer's stamp and has volunteered to prepare two reports that would have otherwise cost me $2,000. The reports are so straightforward that he dashes them off the morning we are due to hand them in.

Leslie Clarke greets us at the door with a smile, and we follow her into her office. Denis does not have an ink pad for his stamp and, used to deference from bureaucrats, sends her off, with his muddleheaded charm, to find one. This one action positions her as functionary. I am impressed by his nerve. I treat her like the queen. She is a friendly, attractive queen, with curly blond hair frosting slowly and a Irish cable-knit pullover. Her eyes are bloodshot, exhausted. I recognize what she is feeling; she is right out at the edge of the complexity she can handle. Mine is one of a score of applications she must finesse through the system or turn down. If she turns something down, she will suffer the anger and pain of the applicant. If she passes something, the anger from the green lobby will hail down on her and jeopardize her job. As a result, her desk is piled high, and work proceeds at a snail's pace.

My anxiety settles into damp scrabbling as Leslie starts puzzling her way through the submissions. Everything seems in order, she says. I have submitted two vast new survey maps of the property at a cost of $7,000, copies of the draft bylaw, the document that waives the minimum lot-frontage requirement, the release from the highways covenant, the new title with the new, big, stricter covenant listed on it, and the approval from the Ministry of Health for septic. Then she gets to Denis's water reports. She blinks, as if to focus better.

My physiological processes start to falter. We have done a fiddle in this report, actually two, but we will probably get away with one. My new well has been bubbling out of the ground at five gallons a minute for the last three weeks. This is, of course, deeply unnerving to those on the island who claim that the water shortage is critical and that therefore all development must be stopped. According to Albert, I have enough water for eighty houses and counting. Not only that, he says, but it is possible that there are a dozen artesian well sites that would pump water from a half-dozen different aquifers on my six acres of meadow, without going up into the forest, where pools of water fed by springs lie open all year long. You see the danger. If all of a sudden there are wells everywhere in which water gushes from the ground like oil, the many proponents of ever stricter water regulation lose the argument that there is no water and that therefore no more development is possible. Albert is plainspoken. He and his father have been divining water for decades, and they believe that there are artesian wells on a nearby ridge with enough water to supply *all the islands*. Not one of the dozen or more citizens groups, water boards, or regional, provincial, and federal governments that have done water studies has ever consulted them.

We dodge the big bullet in Denis's report. The trust requires two pump tests for a subdivision. The old well—in use for seven years with no problem—must be pumped for twelve hours, and measured. The new well must be pumped equally for twelve hours. It is ridiculous to pump an artesian well—$1,000 saved—and Denis has simply said in his report that the old well received a pump test, and it ran fifteen gallons a minute. Which it did. Fifteen years ago.

Leslie's forehead starts to wrinkle. I lean forward over her desk. "What's wrong?" I inquire. She shakes her head and bends over the water tests from the two wells. Then she gets up, photocopies the trust's water requirements, fetches a ruler, and begins to painstakingly make tick marks where my water is good enough and where it is not.

Where it is not is in two biological measurements. The trust requires these to be 0. The problem is that no chemist will write a biological measurement as 0. A finding of nothing is always written "less than 0.1." Leslie, who must have seen this over and over again, shakes her head and puzzles and puzzles. Finally Denis points out the fact that chemical tests can find 0, but biological tests, by their organic nature, cannot. She doesn't accept this. Regulations are regulations.

Leslie admits, after some pondering, that these variations are not critical. What is critical is that the trust requirements for pH and turbidity are not okay. These are aesthetic values, not health-related. The pH at my house well is slightly lower than they like, 6.3 instead of 6.8. At the new well, the turbidity is 1.4 instead of 1. Little use to argue that the rest of the world requires turbidity to be less than 5. Whoever wrote the water regulation for the trust area wanted the highest standards in the world. A covenant must be drafted for both wells and registered on title.

There are tears standing in my eyes now.

Leslie looks at me. "Is this a hardship?"

"Yes," I choke. "At this stage, everything is a hardship."

I glance at Denis, who looks fit to be tied. But we are not finished. She picks up his storm drainage report, reads it, and puts ticks by various paragraphs.

"We usually have a topographic report along with this," she warns.

Denis shrugs. For someone used to massive engineering projects, this is a romp through the Garden of Preposterousness, and Leslie shrugs too.

After the meeting with Leslie, I drive to the new building site to meet Brian Wolfe-Milner, the surveyor. For some of the trades on the island, I am a kind of pet. I am a new member of the club, the victimized-by-conservation club. Yet still she stands! With few exceptions, the people who work on the island, authentic country people, loathe the trust. Brian is collegial with me, and as we stand on the bench where I want to build the house, he describes the pointless, fruitless, expensive exercise he will next perform, of setting pegs on a "safe building site," which I will never use but which nonetheless must be drafted, covenanted, and registered on title.

Brian presents various ways of not setting the posts and thereby saving the money, and we puzzle through them. But in fact there is no workable solution. I have a bill for $500 at my house for the geotech's initial determination. Brian will burn through another $500; the title registration and legal fees will be another $500. All this work is performed to register a site upon which I will not build. And when I build, if I decide another site a few dozen yards to the left is more suitable, I'll need another geotech report, and another $1,500 will be frittered away.

I wake up to a courier banging on the door, carrying a legal agreement from the province, which demands my signature and a check for $29,700, granting me the right to build a house on my lower property. I barely wince. It took them twenty-two months to process a rezoning, which should have taken thirty days, and two years to sell me the density. I decide to pay them in a year or maybe even two. Delays work both ways.

CHAPTER ELEVEN

The Country Is Being Liquidated

When you drive through the singing beauty of the Comanche National Grassland in southern Colorado—on the "lonely road from nowhere to nowhere"—every twenty miles or so, you see, in red paint on a hunk of plywood stuck in the dirt on the other side of the fence:

THIS LAND
NOT FOR SALE
TO THE ARMY

"They want what we got," says rancher Kimmi Clark Lewis. This part of Colorado is Lewis territory. Her family has ranched the Muddy Valley since 1917, and while there is no signage at her gate, everyone knows that Muddy Valley Ranch is not for sale. And if Kimmi Clark Lewis has anything to do with it, not one more acre on any ranch will ever be sold to the government in her county. Nor in any other county either.

Mrs. Lewis strides rather than walks and looks like the kind of blond beauty queen who didn't bother much with looks or clothes after she grew up. She has a decided mouth, a don't-mess-with-me set to her chin and eyes, and she swings aboard a truck as if it's a stallion. Her father was a state senator, and she is descended from a signer of the Declaration of Independence. Public service is fused, therefore, to her sense of self.

Her ranch house is set back from the county road, shaded, utilitarian, inviting. The cliché "neat as a pin" applies, but the home is deeply comfortable. The sitting room is filled with plush, adjustable lounge chairs positioned in a circle, ready for "visiting," which, one gathers, is something Lewis does a lot. She may not have a formal constituency, but in this county, she is the go-to girl.

Portraits of six children in their twenties, all with gleaming dark blond hair and hazel eyes, each an athlete and student in one university or other, sit on the piano.

There are two Western saddles on sawhorses; one was her father's—"the saddle is the last thing a rancher gives away"—and the other was her son's when he won the Mountain States Circuit rodeo finals. Her office, like that of any other mother of six, is tucked behind the kitchen. Overflowing boxes of paper are stacked knee-high.

I had begged an audience with her in a hotel lobby in Denver.

"What's the title of your book?" she barked. "Give me the title of one chapter."

She promised to fit me in between a dinner, where she is being honored as Cattlewoman of the Year, and a trip to New York. So I drive six hours into the farthest southern reach of Colorado, a collection of counties that has been torn apart first by one natural disaster after another and then by land trust officers, leaving broken families, fear, and a sorrow I can't begin to digest.

Lewis talks fast and doesn't give a damn if she's permanently exhausted. President of the Colorado Independent Cattle Growers Association, secretary of Good Neighbor Law, past president of the Southern Colorado Livestock Association, and founder and chairwoman of R-CALF USA, a property-rights association, she ranches cattle on eleven thousand acres and runs a trucking company with one of her sons. Lewis is a tireless speaker and promoter of the rural way of life. Suffice it to say that when one of her enemies at the Colorado Land Trust wanted to talk turkey, he refused to negotiate if she was in the room. She is just that tough.

"This valley is part of my heritage. Ranching is what I grew up to do. We are here to stay. We came with grit, and I intend to stay here with grit, and I hope one of my children has grit enough to ranch, because it takes all of us to save our land." Everyone who meets her hears these four sentences at least once.

After Lewis's husband died of leukemia in 2000, seven years of drought followed. "I had a lot of debt, not to mention having to put six kids through college. The buzzards were flying around, I can tell you. 'Oh, Kimmi, we can take care of all your problems. Just leave it to us.' "

The "buzzards" were land trust officers.

Lewis carries paper copies of her slide show with her, which she distributes to anyone who asks. "I must have passed out thousands right from this briefcase," she says.

The first slide shows that from 1980 to 2008, the United States lost 1.4 million ranches, marking a 53 percent decrease in beef cattle operations, a 90 percent loss in swine operations, an 80 percent loss in dairy operations, and a 31 percent loss in sheep operations.

The second slide shows a steady increase in beef consumption over the past ten years.

The third slide shows the steady rise in beef prices paid by consumers.

The fourth slide shows a steady decrease in the prices received by cattle producers.

The fifth slide shows a 400 percent increase in the number of land trusts in the American heartland.

The sixth slide shows a 400 percent increase in conservation easements in the American heartland.

"I have been studying conservation easements in Colorado for twenty years.[1] If you want to make a lot of money, don't go into producing anything; go into land trusts. From 1950 to 2005, they've boomed. There were fifty-three land trusts in 1950; now there are 1,667. There were 128 easements in 1980, and 6,246 in 2005."

In February 2006, she returned from R-CALF committee meetings in D.C. to a message asking her to call a young man with a map. When she asked her daughter about it, her daughter said, "Oh, Mom, don't worry about it. It's a lunatic. There's always someone out there after you with a crazy story."

"But I got to thinking and I thought maybe I better call this fellow. It turns out he had in his possession a draft land-acquisition map from Fort Carson. It was the regional map of how the army was going to come and take land in southern Colorado starting in '06 and continuing for the next eighteen years, in seven separate phases.

"When I met him, I said 'Young man, I don't have time. If this is just a hoax, I got to git on down the road.'

" 'No, ma'am,' he said. 'Everyone said that you are the lady to talk to and that you would do something about this. I went to that other cattleman's group, and I tried to get them to expose it, and they told me not even to bring that map into the meeting room. 'We don't want to hack off the army.'

"Then he asked, 'Ma'am, are you worried about hacking off the army?'

"I said, 'Young man, I support our troops. We are some of the most patriotic people here. But if this is a real map, we need to know about it.'

"It turned out to be a real map. They were planning to take

2.3 million acres of private land in our valley and beyond. Well, we got on the phone, the newspapers got a hold of it. and it went like wildfire. The *Denver Post* did a story on it, *Rocky Mountain News*, TV came out. When they took the original land, it was 285,000 acres around the Purgatoire River. That was the original maneuver site, which they used twice a year." The land around Lewis's ranch is desert, and the army wants to use it for Middle East war games. But if they already have more than a quarter million acres to train in, why did they need another 2.3 million acres?

"East of the Purgatoire, there was a planned wildlife buffer zone that they were going to take in years one and two, which was going to be supervised by the Nature Conservancy. Then there were eight other tracts that were going to be acquired one way or another over the following sixteen years. It appeared that they were going to take the whole southeast corner of the state; my ranch was included.

"I looked at this map and looked at it, and finally I said to my committee members, 'I'm sure I've seen that map before,' and I thought, well, let's get to looking. And we started digging through boxes. The first thing we found was a memorandum of understanding between the Nature Conservancy and the army dated December 2000, which had been sent to me by a member of my committee. That gave me a clue.

"I had filed the 1997 Wildlands Project map for Colorado, thinking it was so crazy I would never look at it again. We found it, and when I superimposed the army's planned acquisitions map over the Wildlands map, it matched exactly, and both maps planned on taking my ranch. Why? This is productive agricultural land. We feed half a million people out of this county.

"Then someone sent me the press release from the White House Cooperative Conservation Conference on September 9,

2005, outlining the cooperative agreement between the army and the Nature Conservancy. The light broke through the clouds. I thought, why *not* the Nature Conservancy? They own more land than anyone else, why not my ranch?[2] According to their own press release, they are moving to own all the buffer zones around all army training zones. That's the deal they struck with the army and the Bush White House.

"They're everywhere. In Denver, their representative, clean-cut, fresh-faced, looking as innocent and good as the day he was born, takes our representatives out for lunch every single day."

"No kidding," I say. "You mean every week, right?"

"No. Every single day."

When Lewis's husband was dying, the friendly voices on the other end of the phone offered to "take care of" all her problems, and all across the country those voices have rung in many a rancher's ear just as he was hitting hard times. The strategy was so successful, it contributed to the fivefold increase in TNC's assets during the 1990s.

"One afternoon," says Lewis, "I was standing in the prescription line at Walmart, like all us mothers with a lot of children must, and my next-door neighbor, who raised five, was standing next to me, and she turned to me with tears in her eyes and said, 'Kimmi, we have left our ranch just the way it was a hundred years ago, and now they can come in and just take it away from us and leave us with nothing.'

"Their land was right next to the army's land, and was on the list for first acquisition for the Nature Conservancy. They had kept their land pristine—no paved roads, no concrete water tanks—just as it was. And still the Conservancy wanted it.

"This is productive land that provides the tax base in the county. It keeps the schools going; it keeps the community going. There are seventeen thousand ranchers in this area. The land the army wants in three counties feeds a million people. Why are they after us? Why do they need so much land? If the army could prove why they need the land, there are old ranchers here who are so patriotic, they would just give it to them, but there's no rhyme or reason to it. I'm not leaving. Who is to know whether there are mineral rights under the ground which one day will pay off the mortgage? It takes every one of us to keep our land." Lewis repeats her battle cry: "It takes every one of us to keep our land."

The local author of this misfortune is architect Jon Marvel, who tucks himself up in Hailey, Idaho, close to the ski resorts of the very rich, who are his constituency and his cash cow. Marvel went to Phillips Academy in Andover, Massachusetts. His father and grandfather were both prominent Delaware lawyers who ran for the Senate and for governor, respectively. His grandfather was head of the American Bar Association, and his father was ambassador to Denmark under Truman. Marvel is very well connected, therefore, which both served him well and acted as a spur to accomplishment—accomplishment, apparently, no matter what. Marvel came west first as a boy for summers, then as an apparently disdainful back-to-the-lander in the 1970s when he and his wife lived in the family cabin. That cabin was surrounded by public land upon which cattle were grazing, cattle that broke Marvel's fencing and required shooing out of the vegetable patch. Spiritual communing was near impossible after you'd stepped in a giant pile of cow shit. The noise and rustling of the beasts and generally icky industrial nature of cattle grazing infuriated the youthful zealot. He had found his calling. Or rather his not-calling. Marvel devotes his life to getting cows off the range.

Flamboyant, emotionally incontinent, called "immune to reason" by the agriculture weekly *Capital Press* and "Slobodan

Marvel" by an admiring friend, Marvel has been successful beyond his wildest dreams. He has cleared cattle off millions of acres in the West and is considered the most effective advocate for the death of the cattle industry. His stated tactic is to destabilize the cattle industry to the point where ranchers are so miserable that they quit. It works.

"We're creating biological deserts in areas that should be exuberant with life," Marvel says. "If the land could talk, it would be crying."

Compelling stuff to people who never venture farther into the country than the black runs at Ketchum, Vail, and Aspen. Marvel takes susceptible men and women to see water holes in late summer, when the water is low, the streambeds are degraded, and the grasses are desiccated or cropped short, and somehow he convinces them that all streambeds are in this condition. He does not bring them back after a month or two of autumn rain or in the spring, which would lessen the impact considerably. Never mind. Who needs reality? If you think it hard enough, so it must be.

Marvel sues state and federal agencies, files endangered species petitions, appeals grazing plans, bids on state grazing leases, and calls men and women who routinely put in blistering-hard fourteen-hour days outside providing his food "welfare ranchers" and whiners. In fact, he carries a large pink pacifier with him and places it on the table during public meetings when ranchers complain. He has physically threatened BLM officials and routinely shouts at agency employees who refuse to do his bidding. The Bureau of Land Management has twice issued directives that bureaucrats stay away from him, given his verbal abuse and physical intimidation. In 2010 Marvel was cited* for "knowingly and willfully making a false statement," claiming on a grazing lease application that he was acquiring cattle when he was not. Nevertheless,

* Marvel paid the fine rather than go to court.

within the agency, Marvel has many supporters—not surprising, because Marvel's work expands their fiefdom and therefore the security of their income.

Karen Budd-Falen believes that the collaboration over communications strategies and budgets, and the general disregard for boundaries found by the inspector general of the Department of the Interior between agency officials and environmental activists, are routine. Ron Arnold's groundbreaking 1999 book, *Undue Influence: Wealthy Foundations, Grant-Driven Environmental Groups and Zealous Bureaucrats That Control Your Future,* described the perpetual revolving doors between officers of ENGOs like the Wilderness Society, the Sierra Club, World Wildlife Fund, and Natural Resources Defense Council and policy-making positions in the EPA, Fish and Wildlife, Parks, and Interior, providing charts to show how much cross-contamination there in fact is. At lower levels, ENGO employees consult regularly on local ordinances, regulation, and rule writing, acting as de facto policy shops at every level of government. Many in the rural resistance point out that the judiciary is salted through with judges like B. Lynn Winmill, chief judge of the U.S. District Court for Idaho, who is known in ranching country for his many rulings in favor of Jon Marvel.

The abuse Marvel dishes out to ranchers and bureaucrats would not be tolerated from the ranchers. Typically the movement, when it senses opposition among its targets, brings in the troops.

And Marvel's donors celebrate his abuse. "Where Would We Be if Jon Marvel Was Nice?" asks one headline. "True Grit" is the title of a recent *University of Chicago Magazine* puff piece. And environmentalists, of course, praise him to the skies. He is their one-man guerrilla army. His hysterical predictions of doom soften the ground for the big guys when they decide to roll into town and gut a county or two. Which brings us straight back into the all-enveloping arms of the Nature Conservancy.

TNC people know how hated they are in working country, so they use surrogates. In Lewis's neck of the woods, it was the Conservation Fund and Tom Macy. Macy is quoted in David Morine's *Good Dirt* as saying, "We don't like confrontation. We'll just use the U.S. Tax Code to get what we want." At the time, Macy was working for TNC.

The Conservation Fund practices what Macy preaches:

Areas protected: 6,210,000 acres*
Fair market value: $4,056,213,000
Acquisition cost: $2,893,981,000

Impressive, no? But how did this financial wizardry work out for the folks on the other side of the equation?

I am sitting in Bill Lowe's house in Holly, Colorado, population 1,046, about a two-hour drive west of the Comanche National Grassland. It is another profoundly comfortable house, wooden, painted brown, a two-story ranch that almost sags with the life it has housed. You climb the stairs from the garage, decant onto the main floor into one large square room, big kitchen in one corner, a large sitting area, a dining room, a kind of sitting and talking alcove off the kitchen. A giant television broadcasts the latest celebrity catastrophe, Barbara Walters looking spooked and walleyed, providing wildlife watching for country people. The sound is muted.

Bill Lowe stands to greet me. He and his wife are in their seventies. He is tall and straight; she looks like a prairie grandmother—curling gray hair, sharp eyes. Mrs. Lowe exudes comfort: practi-

* As of December 2010. By April 2012, the land trust had preserved 6.7 million acres.

cal intelligence, firm steadiness in the face of all and any disasters. At Mrs. Lowe's invitation, I sink gratefully into the couch.

Denice Kennedy, their pretty forty-two-year-old daughter, is here too. She owns the local restaurant, Porky's Parlor, with her husband, and she interjects often.

"We survived the drought," says her father, "and we'd prospered for ten, fifteen years. Then, we had the worst blizzard in history."

"That was the first . . . ," says Denice.

"We lost $300,000, but that was manageable. We lost fences, we had two hundred sixty calves left, but we had to bring them down into town, into Holly, to the feedlot—it was too muddy to get to them otherwise. Out of six hundred cows, we had two hundred survive; they were all just getting back on their feet. We were replacing the fences . . ."

"And then, March '07 . . ."

"A tornado destroyed the feedlot. That was that. All our irrigation [rolling sprinklers called wheel lines and pivots] was destroyed. Barns were destroyed, but mostly cattle and calves. All the cattle weren't killed, but they were crippled, broken legs, couldn't walk. We had to shoot them, then burn them. We spent three or four weeks shooting and burning in big trenches up on the range. We merchandised some of them for slaughter—we had six hundred due mama cows. We kept sixty. That year was nil." The recitation ends. We are all silent for a moment.

Denice and her brother had wanted their parents to keep the land so they could restore the herd and ranch after their parents retired. So Lowe built a world-class feedlot at immense cost. It met all the new regulations, a model of a clean, humane lot. Three days after the herd was brought down, the tornado hit, taking out the rest of the cows and the entire feedlot.

Almost every house in Holly was destroyed, 160 of them.

People were killed, a child, an old lady. Everyone knew them. And in the days and weeks afterward, up on the range, not too far from town, the men burned the town's livelihood.

"Insurance paid off $120,000. At first, it seemed that we lost a million. Now the figure is more like three million dollars," Lowe concludes matter-of-factly.

"So that's when we got involved with the conservation easements," says Denice. "Dad, you went to the bank, right?"

"Ah, yeah, that we did. 'You do everything right—it's good,' the bank said. We filed for three—they wanted a lot more, but three would get us back on our feet, buy some more cows."

"They wanted gravel easements," says Denice. "This is the place where Colorado gets all their gravel for roads, blacktop, so we hired a geologist to drill and get a report of what's there, and we found a lot. We spent around a hundred thousand dollars, legal fees and whatnot. And then the state said, 'We're not going to honor those easements.'

"We did three. Mine was worth about $1.4 million in gravel; my brother's was worth $1.7 million. We got a tax credit of $375,000 by giving away $1.5 million of land.

"Then Colorado Revenue pulled the tax credits, wouldn't honor them. The state was broke after the '08 crash. But the land trust had gone ahead and filed easements on those three forty-acre lots. We were cooked. We were left with nothing.

"Not only that. My accountant told me I had to use the tax credit. I told him that Colorado Revenue was making people pay back the amount of the tax credit. He said it didn't matter; I had to use them by law."

"A year later," says Bill Lowe, "during the Obama stimulus, a paving company offered us a million dollars for the gravel under one easement. But it's not ours anymore." He gets up, goes to the phone, calls his son to check the facts, then returns.

"There's one hundred and eight landowners who have three hundred easements in Prowers County. Colorado Natural Land Trust won't give us back the land," he says. "They won't release the easements, and I am servicing a debt you can't believe."

He stands up and paces around the room, a man who has ranched 5,600 acres his entire life, has raised a family, and is the pillar on which the county economy rests. At seventy-five, he is very nearly broken.

"There's ranchers out here who have just been ruined over this," says Denice. "They want all this land to go to government. They're cutting ten-thousand-acre easements, letting the prairie dogs run free, which means the range will run to desert in a matter of months and . . ."

Her father finishes: "They'll be using my ranch and others as collateral for loans from China."

One of the studies released by the United Nations in the summer of 2010 claimed that 90 percent of the grasslands of North America were dying.

Since I had just driven through five ranching states, I found this absurd; there was grass for miles in every direction. Furthermore, I saw very few cattle on those grasslands. In fact, I was driving through that perfect world of emptiness and mystery. Abandoned houses and a few struggling strip malls were the only features in an otherwise featureless landscape.

Since the founding of conservation biology at UC San Diego in 1978, more than one million livestock operations had been excised from the range. Those millions of acres should have regenerated by now. Apparently not. Man could destroy the rangelands even when his cattle were taken off the rangelands.

I dug around trying to find a range specialist to talk with. No one would return my calls. I talked to the ranchers who recommended the range biologists, and they said, "Look, they don't want to be public. They are afraid of reprisals, they are afraid of losing ground in their profession, and they don't even want to talk to you. No, not even on background." I was flummoxed. I knew about reprisals, but the worst of mine had been inept death threats delivered by women with a tangential relationship to spelling. Apparently coming forward was not safe.

I called up C. J. Hadley, whose magazine, *Range*, has been the only systematic reporter and analyzer of the devastation caused by the environmental movement in the great west of the country, and she was scattered. She had just undergone a cyberattack, which had corrupted her e-mail. We talked briefly. She had begun her career at Time Inc. as well, at *Sports Illustrated*, then worked at other trades in the Time Inc. stable. *Range* magazine is glossy and slick, with some ferocious reporters, who had, under Hadley's direction, laid out, like a patient on the operating table, the crimes of the patrician movement eating the countryside alive.

"What about the science?" I asked Hadley, who was suspicious of my motives, especially since my provenance—*Time*, *Harper's*, et cetera—should have made me pure green and therefore bent on destroying her. Where did I get my money? At some juncture, she decided I was too inconsequential to worry about and answered a few questions. Like for instance, what's happened to the grasslands?

Most people in the vastly outnumbered fight-back movement dismiss the idea of the science being corrupted. "Well," they say, "so what? It'll just end up as dueling science. My science is bigger than your science." Hadley offered that reason.

"Who judges? The people with the most money, and that is the conservation biology team," Hadley said. "But look," I answered, "we now know what happened to the shut-down forests. Clearly there has been an essential misunderstanding of forest function. Maybe they're wrong about the range too. Were 90 percent of the grasslands of North American really dying? If so, why? It couldn't be cattle. There were no cattle."

She didn't know; she'd get back to me. Six months later, Hadley had apparently got her moxie back, because she published a series of articles on what had happened to the rangelands. In fact, they *were* dying.

In the meantime, I had driven to Montana, then down to Santa Cruz and back up to Seattle, through backcountry most of the way. Different states, same story. Deserted lands, mile after mile after mile. No one on the highways, not even trucks. One broken little hamlet after another. When I got a speeding ticket somewhere in eastern Oregon, I thought, *Well, at least the town will have a little income, God knows it needs every dollar.* What I was looking at was death. Death not just of the little towns but death of millions of acres of rangeland. I came home and told my best friend, Jamie, that it was like driving through *Ghost World*, with wraiths drifting across the fields whispering of what was once all fecundity and life.

"You cannot divorce people from the land," says Allan Savory, a range specialist who is reversing desertification in Zimbabwe. He is on one of several trips he takes to New Mexico every year. "It causes economic and social devastation." Savory is no right-wing, private-property ideologue. He is the founder of something called Holistic Management, which would be embraced by every hippie mother in the universe were he not calling for a fourfold increase of cattle on some ranges and the return of the cowboy. "The range," he says, "covers two-thirds of the earth's [land] surface. In most areas, it is turning or has turned to desert. We have found that increasing the number of cattle four hundred percent increases grass the same amount. Even in the worst drought years, we are finding there is too much grass."

While waiting for Savory to come to New Mexico, I'd discovered that the University of Wyoming had compared the carbon and nitrogen content of grazed and ungrazed land and proved that the carbon and nitrogen levels in grazed land were higher. And several studies out of Nevada had found that when you remove livestock, you remove the caretaker. Neglect and waste take off. Fences are not maintained, ditches and waterways clog, no one controls the weeds, buildings deteriorate without maintenance, and the place falls into ruins.[3]

Grazed land is healthier because cloven-hoofed animals break up the earth, allowing vegetation to root more easily and thereby increasing the carbon content. Savory had come to the same conclusion and on his various ranches had put that science into practice. He had also taken the same drive I had, and seen beyond *Ghost World* to the devastation wrought by forcing cattle off the range.

"Just like a cattleman who has weighed thousands of cows is able to accurately judge any animal's weight as it steps on the scale, so too am I used to judging quickly the 'weight' of any rangeland, after over fifty years of working around the world. Endlessly we cruised through millions of acres of dying ('low-weight') western rangelands. As usual, I kept a sharp eye out for the hordes of cattle causing such degradation, but did not see one. My wife tells me I missed a couple of cows she sighted while I dozed at one point.

"The single most important measure of desertification is, of course, the amount of bare soil between plants. Over our entire journey, these rangelands consistently were . . . around 80 to 95 percent bare soil. Range scientists, and ranchers influenced by them, would judge such rangelands healthy because they have the 'right' species and few 'nonnative' plants, and most tourists would see only open grassland in magnificent scenery.

"Throughout history, land in such condition has never led

to abundance, prosperity, stability, or peace and harmony, but inevitably to increasing man-made droughts and floods, poverty, social breakdown, violence, and collapse of economy and society. Is this going to happen in the United States? It is already happening. One of the clearest indicators of the inevitable cancer of desertification, as such terminal rangeland is called, is the dying of the rich, heartwarming Western ranching and cowboy culture so much a part of the nation's psyche."

You cannot divorce people from the land. Behind all the propaganda from the movement, that divorce is killing the foundational myth of the American West: the bounty of the land, the independence of its people, the strong communities. In the dying grasslands, we can watch the dying of our culture.

The Ehring Bunker

Our two trustees are American. One, George Ehring, is from New Jersey, though he moved to Canada early in his career to act as a legislative aide to Bob Rae, a Socialist premier of Ontario. When in power, that premier, currently head of Canada's Liberal Party, managed to pitch Canada's richest province into an historic decline. When his boss was fired by the people, Mr. Ehring wended his way to Salt Spring.

The other trustee, Christine Torgrimson, is from Montana, where she started something called the Montana Land Reliance in 1978. According to the Reliance's various reports, 59 percent of its claimed successes arose from removing Western ranchers from their lands and locking those lands by easements into perpetual conservation. The people who bought those lands were there to ranch the view and pretend to be cowboys on weekends and in the summers. No hint to be found anywhere in their joyous claims of success as to where their victims were warehoused.

Christine then went on to become an author of the Yellowstone to Yukon Conservation Initiative (Y2Y), one of the monolithic land takings of Agenda 21 and the Biodiversity Treaty. Her "job" was to negotiate with the Indian tribes and bands, whose land Y2Y plans to lock into tight regulation. Y2Y takes most of interior British Columbia, two-thirds of Idaho, and large slices of Montana, Oregon, Washington, the Yukon, et cetera. These lands would be managed in trust fashion, any national or provincial or state authority largely erased and blanket restrictions imposed. In its essentials, Y2Y would confiscate any private landowners' rights. Except the right to pay property taxes.

The first hint of rebellion arose from Christine's decision regarding the Salt Spring Coffee Company. The coffee company is a much-loved institution on the island, claiming the usual fair-trade, organic, shade-grown coffee, which was in fact so good, it became famous across the country. Their success meant that they needed a new roasting facility. So, because "industrial" land on the island is practically nonexistent, they found a site on the main road that was zoned commercial and proceeded to request a zoning variance.

No one said no. But the trust asked, What are you going to build there? If we like it, we might let you. The owners began designing their plant, hiring an architect, and a half-dozen engineers, lawyers, surveyors, planners. Theirs would be an LEED Gold building (the second-highest standard in the Leadership in Energy and Environmental Design certification system), and furthermore, carbon neutral. They would install state-of-the-art scrubbers to reduce any roasting stink to near nothing. It wouldn't be open to the public, so only the employees would use the site—no extra traffic, no commercial activity. Furthermore, the building was very pretty; using the local architectural language, it would definitely improve the look of the broken-down, shingled shacks and bungalows scattered along the main road.

After nearly eighteen months of rigmarole and more than $100,000 spent conforming to every expressed wish, whim, and fantasy of the trustees during a near-endless series of meetings, their application was turned down. Christine's was the deciding vote, and she expressed her reason: "I can't shake this feeling that this is the wrong location . . . " she said. "I've seen [sprawl], I'm sick of it."

The island erupted in anger, and resentment poured like lava from the citizenry. The Islands Trust is fond of suing people, generally middle- and working-class islanders who contravene one of its many, many, many regulation sets, always people who cannot afford the $150,000 it costs to challenge the trust in court. On a smaller, sister island, a lawyer who had made his career working for the disabled started a petition demanding a core review of the trust. More than a thousand islanders signed the petition, and in that petition, hundreds expressed their distrust of the trust and their frank hatred and fury at the harm it had done to their islands, their community, and their neighbors.[1]

That summer, we held a rally in the park on Salt Spring, on the Fourth of July, a coincidence we exploited by calling it our Independence Day and the Salt Spring Coffee Party. We dumped coffee into the harbor. Forty naked or near-naked farmers and tradesmen rode into town on their tractors, trucks, and excavators to join us. Larry Campbell, a Vancouver coroner whose career had become the basis of a famous TV drama series, which propelled him to mayor of Vancouver, then to the Senate, spoke of his dismay. We got national press and television and embarrassed Christine and George and their supporters.

A resistance was forming on Salt Spring. Anarchic, impossible to control, and entirely amateur, we wobbled toward the goal of kicking the extreme elements out of our politics. Several men and women focused on the corruption of the democratic process typical of conservation government, and one woman, Jill Treewater, started videotaping every single meeting just in case the trustees, drunk with power as they appeared to be, tried to openly subvert that process. She filed those meetings on the Internet for everyone to watch. The trustees protested the videotaping and were fought. Jill won.

One of our sister islands had incorporated itself a few years before and as a result were governed by a typical town council, with the trust on the council but not in the majority, which meant their power was substantially diluted. One of those councilors was asked to come over and talk to us about how having a town council worked in the trust area. I know such a thing seems elementary, but the trust had convinced islanders that without its full and total control, developers would run rampant and turn the place into an awful gimcracky tourism circus. But Bowen Island had managed to forestall that turn of events handily.

George Ehring decided that no outside politician was allowed to talk to Salt Spring Islanders without his approval, and he had forced through a resolution at the trust saying just that. Even the slumbering middle class found that exercise in authoritarianism egregious.

And then the death blow, or rather a necessarily unpredictable, utterly surprising transformational event occurred. An anonymous woman who called herself V and described herself as a grandmother and pensioner started a blog called Salt Spring Folly. Her first posting was that famous Internet video of Hitler berating his officers,[2] Hitler's words in this case being replaced by George Ehring describing how he had destroyed the economy of the island, turned the island into a park for civil service pensioners and rich women, and furthermore would destroy his critics.[3]

The whole island fell about laughing, and the countdown to the election began.

CHAPTER TWELVE

High Noon in Owyhee: Rebellion Begins

GRAMPA: And I don't give a hoot in a hollow if they's oranges and grapes crowdin' a fella outa bed even, I ain't a-going to California! [Picking up some dirt] This here's my country. I b'long here. [Looking at the dirt] It ain't no good—[after a pause]—but it's mine.

—Nunnally Johnson, screenplay for *The Grapes of Wrath* (1940)

I am sitting in Fred Kelly Grant's dining room in Nampa, Idaho, watching the massive television screen that dominates the neighboring sitting room. It is pivoted toward us and tuned, as it generally is, to C-SPAN. We are watching Harry Reid spin his magic after the November 2010 trouncing, and I find myself nodding like a robot. I jerk myself out of it and remark on his ability to hypnotize.

"So conciliatory, it's as if angels are flying out of his mouth," says Grant. Grant, of course, recognizes a colleague, being no mean hypnotist himself. His South Carolinian lilt has magicked him through courtrooms from Chicago to Baltimore to D.C. to Idaho, through hearing rooms in Texas, Montana, California, and a dozen other states, then onward to an endless succession of conferences, teaching sessions, and confabs.

Most people in the rural resistance are Democrats turned Republicans turned seriously pissed off, and Grant is that to the power of ten. He was raised a southern Democrat who as a kid used to ride the county with his uncle the high sheriff; his late wife's coal-mining family were devout Democrats, who "thought John L. Lewis was the Second Coming of the Savior." He never imagined he would quit the party. He went to the (then liberal) University of Chicago law school, was a U.S. attorney in Baltimore—"you had to be a Democrat if you wanted to do anything in politics in Baltimore"—then worked for two Democratic governors. So turning Republican ripped a tear in the fabric of his universe, but there was more to come when he discovered that big business had bought in and was supporting the draconian regulation that promised to shut down everything. "When we fought ICBEMP, Boise Cascade [a major forestry corporation] should have been in that fight, but they supported the regulation, since it would eliminate all their local competitors." So he left the Republican Party, too.

"I think they're both awful," he says. "I'm a libertarian, and independent. The pernicious Food Safety Act[1] they're trying to pass in this lame-duck session is supported by Monsanto, which is great for them, because it'll put the mom-and-pop farms out of business. The farm-state senators supported it because Big Ag is where they get their money."

His is a typical political progression for anyone in the rural resistance, which, if it could be said to have a charismatic figure, has one in Grant—that is, if a plump seventy-seven-year-old lawyer with a habit of car accidents, heart attacks, cancer scares, and general disasters trailing him like a herd of minor-league demons can be said to be a charismatic figure. But I know he is the leader, because his is the only solution that makes sense. I have listened to him for several days and can't find a flaw in his reasoning.

His house is a mess, situated in an untended garden on an untended side street in a suburb of Boise. His wife—who one assumes bore with his decidedly absent organizational skills and his former two-fisted drinking—has gone on to her reward. There is still evidence of her around—a china cabinet, the front of which is now crammed with sports memorabilia, and a charming kitchen wherein every surface is stacked with office supplies. Her prints hang awry on the walls. Her splendid dining room table is now a giant desk, in the center of which, as a metaphor for Fred Kelly Grant's life, sits a tray filled with cell-phone chargers, because he loses a charger a day.

His children work with him. His daughter-in-law Staci attempts to impose order—"for a while we had to sit with him in the evenings so he would remember to eat." His son Jon's office is in what one assumes was the younger man's childhood bedroom, the walls now hung salon-style with photos and caricatures of Grant's favorite sports stars. Their attitude is one of bemused irritation, but I notice that when he asks, they jump. Everyone does. He's that good.

After a few years with a white-shoe law firm in Chicago, his wife's rare blood disease took the couple to Johns Hopkins, and in Baltimore Grant was made the youngest-ever assistant U.S. attorney in the district. Two years later, he became chief of the Organized Crime Division. After five years, he resigned and started defending the people he had been putting away. He never lost a first-degree murder case and never got a verdict returned that was higher than he'd offered. "Prosecuting organized crime and then defending people in it prepared me to deal with the federal government. The systems are much the same." This line invariably gets him a laugh, until he goes on to illustrate how he arrived at his assessment.

His wife's illness brought them home to Idaho, where he took a job as counsel to a succession of governors, which meant, at first, a 90 percent pay cut. He cobbled together a living, raised three strong, attractive children, and then she died, but not before Grant had found his life's mission. Grant always prefaces this story with an anecdote of his attending a real-property class at the University of Chicago. The professor settled on Grant to elucidate a dense paragraph from English common law, and Grant had been so thoroughly humiliated, he'd sworn to his best friend afterward that he would never, ever practice property law.

Then one day a county commissioner pal called from the Jordan Valley, seventy miles south of Nampa. Owyhee County, with five million acres and about ten thousand people—to get the density, think one house per square mile—was about to have its economy drawn down. Bruce Babbitt and Bill Clinton had decided that 40 percent of the ranchers were to be taken off the range. The only business in Owyhee is ranching.

"Idaho Watersheds [now Western Watersheds] brought in people to specifically focus on Owyhee County—the Biodiversity Center, National Defenders of Wildlife, Rivers United, and eventually the big guns, the Wilderness Society, the Sierra Club, and the Nature Conservancy. They figured it was going to be easy. Pretty much everyone in the nation thought the ranchers of Owyhee were rabble-rousers, so no sympathy would be shown. Which meant, we get them out first, and then every other ranching community we target will follow easily," says Grant.

Owyhee County was one of the loci of the Sagebrush Rebellion, which sprang from Jupiter's forehead in the 1970s and 1980s to contest the sequestration of public lands for wilderness in the West, especially those lands traditionally used by ranchers to run cattle. Because much of the West—roughly 60 percent—is public land and cattle need large acreages in arid regions to browse, ranchers with two thousand acres, say, would lease another five thousand or fifteen thousand acres from the Bureau of

Land Management. Those leases were passed down from father to son for generations and had property and water rights vested in them. Many Western historians believe that those public lands were meant to be sold to individual ranchers eventually, but in an early manifestation of eastern class-based distrust of western roughnecks, that intent was thwarted over and over again.[2] The entire ranching economy—the founding economy of the West, the mystique of the cowboy, the freewheeling, ecstatic *soul* of America—had been based on those leases.

And then in the 1970s, the leases began to be pulled. The first reaction was fury, then rebellion, then more fury. And then, as haggling with the federal government turned into the kind of negotiation that one would experience with the Mob over, say, having to pay for garbage removal in lower Manhattan, frustration and despair. A few fought on, notably Wayne Hage, a Nevada rancher who discovered Forest Service men packing guns on his allotment one morning in 1979. They informed him they were taking his water. Thirty years later—but not until his wife had died, his second wife had died, and he had died—his estate was awarded $13 million in damages in his decades-long suit against the Department of Interior. His children carry on the tradition. And the fight, because of course the government appealed.

Hage was at Grant's first meeting in Owyhee County, along with other ranchers. Grant recalls: "Jordan Valley was holding the National Lands Council conference, so I drove up there in January, in a snowstorm, thinking what the *hell* am I doing? Karen Budd-Falen spoke and Wayne Hage, and I wasn't buying any of it. They were going to assert county supremacy. The county got to call the shots, they claimed, and if your custom and culture were in danger, why, the feds had to back off. I said, 'That's the most unconstitutional thing I've ever heard. You can try it if you like, but you can do it without me.'

"I told them that if the federal government decides to take your cows off the land, they're going off the land. There's nothing you can do about it."

But the commissioner persisted, so Grant said he'd take a look at some statutes. "I called a friend in D.C. to find out what the Bureau of Land Management was and how it was run and governed. I had only a whiff of its meaning, because my wife's ranching family in southern Idaho used to call it 'the g'damned BLM,' which I thought was its actual name. My friend directed me to the Federal Land Policy [and Management] Act.

"So I was sitting around here one night and started reading FLPMA [pronounced FLIP-ma], which governs the rangelands of the West. Buried in it was a clause which read, 'The Secretary of the Interior shall coordinate with local government . . .' And the next seven sections defined what the responsibilities of the federal government were with regard to local government, which are: 1. Make that government know ahead of time that they're going to adopt a plan. 2. Develop a plan. 3. See whether there are inconsistencies with the local plan. And 4. If there are inconsistencies, they have to use *all practicable means* to make their federal plan, rule, policy, or management action *consistent with the local plan*."

Grant was set back on his heels. In his entire career, he says, nothing had surprised him so much. It was as if the Founders had reached through time and reordered his universe.

"I could not believe that the Congress had let that get into the legislation. In fact, I found out later on that Senator [Bob] Packwood had inserted it, in part so that local environmental groups could force change, but also because the rural communities of the West depended utterly on those public lands. He informed the agencies that they couldn't run roughshod over local people. But the root and rationale came from the first Continental Congresses. The Founders knew that county government, local government, is the most important unit of government. The first decision made at the First Continental Congress was whether a representative from a county in Georgia could sit in, even if his state had not yet decided to formally participate. They knew how important counties are. Who talks ten minutes with a senator, a congressman, even their state representative? But every adult with a working brain has talked for ten minutes with their local mayor or councilman. And every law that Congress passes is implemented at the local level. This clause meant that every regulation had to dovetail *without exception* with what local people had decided they wanted in their county. It was a revelation."

When he recovered from his shock, he called up a friend at Justice in D.C. to check his interpretation. "A few days later, they got back, and said, yup, it's there, but as far as we can tell, no local government has ever used it.

"Using coordination, I figured I could only hold back the inevitable for a couple of years using administrative reviews, and even that was going to be really hard. I told the ranchers that they'd have to get really involved in planning, but that at the end of those two years, we'd lose, and the government was going to take their lands. At which point they were going to have to find another way of making a living.

"Ten years later, they're not only still there, but some of them have added rangeland to their ranches. And the Sierra Club sits at the table with the tribes, the air force,[3] the ranchers, and recreational users, and together we all negotiate sound wilderness policy, range use and management, restoration, and leasing. The Congress cites us as the *only* success story in the whole United States.[4] We are the only county in the entire United States which has come to an agreement."

In fact, bringing peace to the Owyhees meant that Grant had to get a bill passed in Congress, but once that was done, it worked; longtime antagonists even became friends. Wilderness in the Owyhees is still being added, but without damaging the livelihoods of the people living in Owyhee County. Likewise, range restoration no longer consists of dictates handed down from on high. Local ranchers share their deep knowledge with range scientists, and progress inches forward in fits and starts. The significance of the coordination requirement is that Big Green works mainly through lobbying at the federal and state or even supranational levels and deliberately overrides mere local consideration. Coordination says no, the way a place is run must reflect on the well-being of the people who live and work in that place.

In 2008, the fellows at PERC celebrated a new vision for the great middle of the country. The "vision" was that of two visiting professors at Princeton, Frank and Deborah Popper, who had decided in 1987—based on their strong feelings—that a "buffalo commons" should be created because the settlement of the Great Plains was "the longest-running agricultural and environmental miscalculation in American history."[5] PERC pointed out that their prediction had turned out to be remarkably prescient, that the plains *have* gradually depopulated—fewer people living on them than since frontier days—and a quarter of a million buffalo now roam throughout the West. Was the Popper vision correct? asked PERC. Yes, yes it was. In 2008, people still thought a green economy filled with ranchettes and tourism and "alternative choices" would provide an alternative to mining, logging, farming, and ranching the raw materials of life. iPads, after all, come from the ether and the beneficence of the deity Steve Jobs and were not constructed from raw materials and sweat, or at least not American raw materials and American sweat. Same with red meat, chicken, wheat, vegetables, and fiber. The Poppers did not ask what happened to the millions vanished from the Great Plains. Nor did they take credit or responsibility for the subsequent devastation. Their idea, however, propelled both to Princeton and no little amount of fame.

By 2000, Jon Marvel had teamed up with Andy Stahl, who had stickhandled the spotted owl shutdown of the western forests when with the Sierra Club. Together they decided to start something called RangeNet. Its aim was to get all cattle off public land, no matter what. The RangeNet Declaration was filled with the usual sticky, bathetic, and largely meaningless enviro language, terms like *wild*, *diverse*, *vulnerable*, *complex*, *restore*, and *American heritage* and terms like *livestock industry*, *destructive*, *damage*, and *massive subsidies required* for their class enemies. The "proof" of their stated need to clear animals off the range so the range could recover was a collection of accusations that, if true, would raise anyone's blood pressure. The shimmering goal of a "healthy, sustainable economy in the American West" was held out, with no real-world idea provided of how to achieve this new green economy—tourism, as if anyone would have the money to travel after they'd gutted the economy.

It's time to pause for a moment and consider that the green economy is a mirage and has never ever, ever shown itself to work other than in the imaginations of privileged, half-educated men like Jon Marvel and Andy Stahl, and in the minds of the thousands of others who have used their class privilege to damage the lives of those far less fortunate. Neither Marvel, Stahl, Robert F. Kennedy Jr., Teresa Heinz Kerry, nor her dreadful husband could handle one single day of labor routinely performed by a rancher or a farm wife. Along with their pet corporate bosses and politicians, they've snowed us with hysterical science, celebrity connections, and the awesome snob appeal of stinking-rich private foundations; they are an onerous tax on the system. They certainly do not see the devastation they have caused. I can. I think that, overall, their actions have diminished working- and middle-class incomes by 20 percent over the last twenty years. And for men and women in working country, it's 40 percent.

A year after RangeNet had stuck up its little princessy head, a rancher and former BLM area manager delivered a dose of reality, sentence by awful sentence, using actual evidence rather than virtual "science." First of all, said Ed Depaoli, leased rangelands were healthier than they have ever been. The use of public land allotments, when managed with evidence-based management—rather than panic—is fine. Left to its mandate, allowed to do its job rather than fight lawsuit after lawsuit, the BLM uses up-to-date science, on-the-ground testing, and generational knowledge of the specific region to hammer out with ranchers the best use of the land. Depaoli is too polite to state the obvious, that Jon Marvel had no earthly idea of how real life works, that most people act with goodwill, and that expensive educations are more often than not put to productive use rather than the destruction of a way of life. Depaoli did point out that great strides over the past decades had been made in stream and riparian stewardship and that fencing riparian areas, as Marvel and his cohort demanded, meant that invasive weeds took hold and spread metastatically onto pristine rangeland.

As for the claim that the range improved when cattle were removed, Depaoli cited the work of J. Wayne Burkhardt, a professor of range management at the University of Nevada who spent years studying the effect of long-term livestock removal using large enclosures and found that "there are virtually no differences in the plant species which occupy the grazed and protected areas. . . . Fifty-six years of protection from larger animal grazing has resulted in identical plant communities."[6]

As far as wildlife is concerned, wildlife increases where animals are ranched. In 1978 John L. Schmidt and Douglas L. Gilbert compared the following wildlife population counts:[7]

	1900	1978
Mule deer	500,000	3,000,000
Elk	41,000	1,000,000
Antelope	12,000	1,000,000
Bighorn	2,000	45,000

Driving into Jordan Valley is like driving into the past, and I don't mean the 1950s; I mean the town looks the way every western town must have looked a hundred years ago. The rodeo ring is a tantalizing maze of weathered gray boards, no neon, no advertising in sight except a couple of faded tin signs; the houses are modest, and the roads haven't been bulldozed into straight lines. The gas station profile is so muted that it actually blends into the town center, which is essentially a crosswalk with a feed store. Three miles to the left after you reach a T-junction, down a gentled country road, past one ranch, then another, the Lowry home is third, a collection of saltboxes painted dark red with white trim, so simple the eye slides over them to the fields and forest beyond. There is not a scrap of debris or weed in the yard. The people who live here are so virtuous that everything has been tucked into its proper place and nothing is wasted.

Inside it is maybe 1965, and the two couples who greet me are so familiar, it is as if we'd known each other for years. The wife of the older couple is tall and rangy, a strong eighty-something, with capable hands, her hair still dark and wavy. Her husband, Bill, is shorter, with the shiny face of a melanoma patient, but he is sunny and, like near every genuine country person I've ever known, wry and self-mocking. His son, Tim, has a handlebar mustache, the drooping ends of which are waxed. Other than that oddity, he is conventionally handsome, strong, weathered. His wife, a teacher, has the comfortable, intelligent face that any smart eleven-year-old would be relieved to see every morning.

They are making lunch together, as they do most days, the men coming in from the range. Hamburgers and salad. Fried potatoes, ice cream and chocolate sauce, cookies and coffee. Bill is almost ninety, but they can't afford help, so he still works outside most days, all day.

It takes a while to draw out of them the story of how they put together their ranch.

"Oh, yeah, okay," says Bill, finally, a little embarrassed. "We worked at another ranch for twenty years, saving half of the money for the ranch out of our three-hundred-dollar-a-month salary. Well, more at first. I used to give her fifty dollars a month housekeeping money, and then when the two boys came along, we upped it to a hundred fifty. The ranch owner kept the balance and paid us interest on it. So when we came to buy the ranch, we couldn't afford contiguous parcels. But we have five thousand acres, lease nineteen hundred acres of BLM and state land now, and run about five hundred fifty mama cows."

The Lowry land, like much of Owyhee County, is full of juniper, a mean, twisty, shrubby dwarf cedar that can take over a pasture and desertify it in a month of Sundays. Jon Marvel christened the Owyhee an "endangered juniper montane forest," which would be risible if so many people didn't believe it. When juniper encroaches, as it will, like a crazed shopper on Black Friday, it destroys forage, including sagegrass and sage grouse habitat. The understory disappears. Bird life, including the grouse, vanishes as juniper creeps over the range. Cows searching desperately for forage disappear into it like smoke in the air. These days, the Lowrys have one field where they are allowed to graze their cattle for thirty days, but the juniper is so thick, the cows vanish, and it can take six months to find them and pull them all out. So Bill and Tim put the cattle in for a few days but restrict them to the front of the pasture so they don't get lost. The cattle graze the front of the pasture down; then the BLM comes in and says it's overgrazed and fines them. The BLM says the grass must be six inches high. Ranchers often have to tell ruler-enhanced and seemingly witless BLM functionaries that the grass isn't six inches high because it's *new spring grass* and still growing. Stream depth is measured, the minimum speed of stream flow is determined, and they take out a ruler to measure the average height of grass over five million acres of deep canyon and high mountain desert at the end of grazing season—tactics meant to make Idaho cattle-free by '93.

"It was '87 when the big wrecks started here," says Bill. "The enviros were getting started before that, but in '87, the agencies [the Bureau of Land Management, the Forest Service, Interior, Fish and Wildlife] started doing their dirty work, and things got very bad. First they increased standards and guidelines on grazing. If you are going to do things outdoors, you have to use tools. But wilderness study areas had been established in '81 and the agencies started interpreting the act at the behest of the enviros, so that you couldn't do anything in a WSA with tools. We were being set up to fail. You have to meet this standard, but you can't use the tools that range science requires. If you don't meet the standards you get cut."

The Lowrys are considered to be one of the three most environmentally conscious ranching families in the valley. They agreed to take their original twenty-thousand-acre allotment and divide it in half with a fence to create a deferred-grazing arrangement, leaving one pasture fallow every year. But the allotment was in a wilderness study area, and that meant no tools. And a fence would detract from wilderness characteristics, so that was out too.

"We fought that for a number of years and even got our congresspeople involved. This pushed things up to 1990. Congress was supposed to act on these WSAs in '92 to determine whether they would be designated as wilderness.

"The national BLM director issued a directive to state BLM managers that any improvements in a WSA had to be completed by 1991. We'd been fighting and fighting and trying to get the tools—generators, power tools, ATVs—to get this done for years so we could stay in business, being refused, being refused, being refused. Finally, in September of 1990, we thought we were done because the cutoff date for any improvements was September 30th. Our local BLM manager met us on our way down from the South Fork allotment on the 26th and told us they'd finally decided to let us put that fence in. That gave us four days to put in two miles of fence. In wilderness. That was real peachy of them, we thought. So, okay, four days; it's tight but we'd get it done. No mechanized equipment, though, too destructive of the wilderness. And it had to be complete by the date, or the whole thing would be ripped out. Even if we'd strung two out of the three wires.

"The county turned out—kids, wives, everyone came—and the two miles of fence, that's eleven thousand feet, was done.

"So that following year, we got to use the system. But then enviro groups started protesting the fence and filing litigation. The BLM settled with them, took their fire crews, and tore the whole fence out, and you can bet they used mechanized equipment. They left wads of crap in the wilderness, tire tracks, wasted the material, jerked the fence right out of the ground.

"I read the complaint brief—that fence interfered with their religious experience out there in the wilderness. I don't think any of them had ever been there. They couldn't commune; it ruined their spiritual experience. We had two-thirds of our allotment taken away from us, twenty thousand acres. Half our income, lost."

• • •

The best kind of war is an all-out assault so cruel and pointless that people eventually just give up. Picking off one rural family business after another, especially people with little access to the kind of money and power habitually wielded by the movement, is so easy that all that's required is a wireless connection and a lack of conscience.

On the Colorado Front Range in the early 2000s, thousands of farms were put out of business one by one because of lack of water. The water had been sent to South Dakota for the threatened whooping crane.

My mind's eye pictures a hundred-acre field of ankle-deep water in which the cranes (mine are pink) frolic happily. But the whooping crane doesn't actually live in the Dakotas. A couple of years after 250,000 acres of the most productive farmland in the Rocky Mountain West was mothballed on its behalf, people realized that the crane was just passing through—two weeks maximum. A typical urbanist would say, so what? The crane needed the water. We need the crane.

Coloradan Chuck Leaf, a longtime hydrologist with the Forest Service, continued a study he'd started in its employ after he left in disgust, and he discovered that if the forests on the Front Range had been managed even a little, *even mismanaged*, there would have been enough water for the crane, Denver, Aurora, all the new subdivisions around Denver, *and* the farms.

The turning of the screw gets down to millimeters, and I mean real millimeters, not metaphorical millimeters. In northern Montana, an irrigation district begun by a group of ranchers in the 1920s has been in a decade-long battle with Fish and Game over the endangered bull trout, which was released into the irrigation lake by Fish and Game. Despite the fact that both lake and ditches are man-made, the bull trout must survive in that water system.[8]

• • •

It really is all about the water. It's not a coincidence that Marvel calls his outfit Western Watersheds.[9] Control the water, and you control everything. Water for America's Great Plains, its ranch lands, farmlands, forests, and cities comes off the snowpack of the Rockies. The Missouri runs eastward across the Great Plains. The Colorado snakes west, forming the Grand Canyon, and it is the major watercourse for the Southwest. The North and South Platte rivers start in the eastern Rockies, the twin tributaries joining in Nebraska and turning in a great arc to water the plains before draining into the Missouri. The Columbia is the life giver in the Northwest. Each river system is mythic, replete with story, history, and use. Dominate those waters, to the extent of mandating the depth and speed of irrigation and runoff ditches, and you dominate the life of the West.

Formally, legislatively, a host of corridors is being established over the continent. The Western Climate Initiative draws seven western states and two Canadian provinces together to effect cap and trade without the necessity of any law being passed by Congress or Parliament. It is an extrajudicial administrative body that does whatever it likes. The WCI planned to start auctioning carbon credits in the summer of 2011, until the Chicago carbon exchange collapsed, but like water flowing around a rock in the middle of the creek, if not stopped, one way or another, the WCI plans to be taxing and capping every economic activity in those states by 2015. The plans overlie each other in a ten-dimensional choke hold on the productive lands of North America.

In November 2008, the Department of Energy announced the designation of more than six thousand miles of "energy transport corridors" in the eleven western states. Animal migration or wildlife corridors from which humans are excised proliferate. Y2Y,[10] or Yellowstone to Yukon, takes 5,120,000 acres of Montana, Idaho, Wyoming, Oregon, Washington, British Columbia, and Alberta. Man will no longer be allowed to access resources in those 5,120,000 acres. After that land is locked down, activists will double the corridor size with buffers the size of Texas. Property taxes will be reduced as development rights are stripped from those lands, which means that money for local elder, school, and health services will be gutted.

The National Wildlife Federation, along with twenty other ENGOs, is working to tie up eighty million acres on the eastern seaboard in a plan known as Algonquin to Adirondacks, or A2A. Baja to Bering (B2B) controls the use of the waters off the West Coast of North America from Mexico's peninsular tip right up the Alaska coast and curving around the North Slope. America's western waters will become either off-limits to Americans or strictly limited. These transborder regions will be supervised by unelected multinational committees with no accountability to local citizenry, nor direct accountability to state or national governments.

The 25-million-acre wilderness set aside by the Northern Rockies Ecosystem Protection Act, or NREPA, bleeds into Y2Y and locks down those additional millions of acres in Idaho, Montana, Wyoming, and Oregon. "Travel management" plans in the National Forest system tighten access in every rural county, closing roads into the forest, turning it to wilderness, making it difficult if not impossible to fight fire, restricting access of local people for firewood, recreation, and hunting, destroying cultural activities in place for hundreds of years. Control of water rights, control of mining rights, control of private-property inholdings through conservation easements, like the one I have on my land, tighten use and access further. The Biodiversity Treaty that shocked the Senate so much in 1995 it voted 95–0 against the bill, has been effected piecemeal in any case, without transparency, legislation, or even notice.

Every advance that tightens control on land reduces productive use. The cost of raw materials rises, and the value of land pitches downward. If we accept the indisputable fact that set-aside land is now dying, then this monster, this hydra-headed slouching beast, is not just the apparition of doom; it is Doom.

The only people who win in this game are the ones making and executing the plans. Ron Arnold[11] estimates their number at only twenty-five thousand, outside of those activists embedded in the federal and state agencies and the legal system. Twenty-five thousand doesn't sound like many people, unless each of these people represents four or five different organizations with the requisite noble-sounding names. On my last trip south, I sat across the table on the ferry from an attractive young man, politely alternative with long, clean, sun-kissed hair and beaded bracelets, fixedly working on his laptop. He stopped only to call his girlfriend to establish which of their several organizations' letterhead to use in order to stop the thing that he wanted to stop.

He, at least, had a job.

Buffering Paradise

At home, we have settled into the ultragreen house. There is much to be said on its behalf. The walls are rammed earth,[1] and two feet thick. The uneven tones of a dirt wall dyed red with iron oxide and lifted with white cement have produced porphyry, a deep rose that casts an ancient flush, as if the house were pitched up here from Egyptian prehistory and set in a symphony of intense greens, decidedly odd in the hypermodern Pacific Northwest.

The difficulties of matching the precision required by modern building to an ancient building system that is always off that crucial quarter inch were impressive. Our engineer worked hard and long to make sure the house was tied together and rock solid, as it were, and despite its twenty tons or so, if the mother of all tornados hit, the pink house would probably lumber off to Oz intact. It feels at once both lost in time and ridiculously solid, as if the house had decided, once it was fully born, to stand for several millennia, come what may.

Rammed-earth walls are built in lifts. Slurry is poured into plywood forms, then hand-rammed into a solid. The wall is softly various; in fact there are no surfaces machined to the dead flatness of Sheetrock; the shades of difference in concrete floors, birch, alder, and fir trims, doors and windows mean all look alive.

Its radiant heat is also somewhat preternatural. Delivered through the concrete floor by geothermal, its warming and cooling are superior to any I've experienced. Imperceptibly it softens you; you aren't aware of the temperature changes; it is subtle but immensely luxurious. The pink house has three times the volume of my cottage but costs the same to heat, once you've eaten the cost of installing this green refinement—in this case, sixty large. It'll pay for itself in twenty years.

I picked my way through the LEED building handbook. I have followed the green-building template, and if I could have stood another official with his hand out, I would have had the house certified. The house could receive a "healthy house" certification too: no solvents, glues, drywall, formaldehyde, or paints were used; therefore there's no new-house off-gassing. Let's put it this way: of course some green is valuable, and most of its sentiments praiseworthy, but all of it is expensive and complex. Nonetheless, mine is a soulful house—probably because it ate my soul—and I am proud of it.

A few months after we surmount the final hurdles, I get wind that the rest of my property is about to be locked down in its entirety under the new islands-wide riparian regulations. By the autumn, if the trust gets its way, I will have to get a permit and hire a registered environmental professional (REP) in order to turn over a shovel of dirt. On my 16.5 acres, configured as it is with a creek cutting through on the diagonal, I won't be able to spread a load of gravel, plant a tree, dig a new garden bed, transplant a fern, erect an arbor, or cut vegetation to clear a forest path. And from now on, I must plant only indigenous shrubs—no lawns. Unless I pay that REP $2,500 and go another round with the Islands Trust.

A map is published showing that these blanket restrictions will cover 65 percent of my island, which means most property owners will be affected. Those with oceanfront property will be locked down with another set of regulations, equally restrictive, once the riparian rules are in place. The excuse is the protection of water. A hundred-foot buffer will now be in effect for all watercourses and bodies of water, man-made or natural, seasonal or not, including runoff ditches along the side of every road. The evidence presented to advance this agenda is rife with the following locutions: *may be, might, can be, likely to.* Once in place, houses within a hundred feet of any water body will be designated *legal nonconforming*,[2] meaning that, if they burn down, say, it is possible that they would not be allowed to be rebuilt.

To burst out of a growing claustrophobia, I start driving to Seattle. As advanced as the agenda is in my region, south of me it is demonic in its complexity and progress. I am an admiring student of the battle. The Seattle area is home to many senior scientists who planned to live near lake, ocean, creek, or pond in retirement and found themselves hemmed in by restrictions even more constrictive than those on my island. Curious, they dug into the specific science behind the blanket property confiscation and published their findings. A few proceeded to sue bureaucrats for the disastrous results created by that corruption of science. Won. Crashed liability insurance for those bureaucracies. Burdened state risk pools. Watched while that bureaucracy ran up fines in the hundreds of millions.

Forgive me for taking pleasure in this.

Finally they brought together 150 equally outraged scientists to found the Environmental Sciences Independent Peer Review Institute in D.C. to formally investigate the assertions of conservation biologists.

PBS's 2009 *Poisoned Waters*, a two-hour dirge hosted by veteran journalist Hedrick Smith, encouraged the rapid spread of riparian regulation, which is now or is about to be in place in every jurisdiction in North America. With doom-drenched strings and drums and noble poses struck against magnificent waterscapes, Smith uses Maryland's Chesapeake Bay, outfall for the largest, most industrialized estuary in the United States, and Seattle's relatively pristine Puget Sound to prosecute industry, population growth, and general human malfeasance for the death of sea life. It's not surprising that there are dead zones in Chesapeake Bay.

Evidence is produced (more doomy music) that promises the imminent death of Puget Sound. Smith's interview subjects state that this was largely Boeing's fault.

Poisoned Waters explains that much of the Seattle-area pollution happened thirty years ago, before the dangers of PCBs were fully discovered and they were phased out. The documentary charges that PCBs, endocrine disruptors, overfishing, industrial sewage outfalls, storm-water runoffs, and the bewildering array of new chemicals created by industry every year are overburdening nature's filter—the ocean. No one knows the effects of the new chemicals, and especially the cumulative effects. Frogs are bisexed, as are fish.[3] Killer whales in Puget Sound are failing to thrive, and some are dying young.

"Seven million Americans *might be* [my emphasis] sickening from water contamination every year," intones one of Hedrick Smith's interviewees, adding that this is at the high end of his guesstimate. Still, one thinks, less than 1 percent *possibly* sickening is a good deal better than one in four children before the age of five dying from water contamination, as is true in much of the developing world. Smith doesn't mention that we already spend hundreds of billions every year to ameliorate industrial pollution. As fast as new chemicals are discovered to cause harm, new technologies to abate that harm are put into use.* The very definition of adaptive management is "scurrying to catch up to the latest catastrophe," and frankly we have come pretty far using that method.

Poisoned Waters perfectly illustrates what is wrong with almost all environmental reporting. It slides over the fact that the Duwamish River, Seattle's industrial waterway, was declared a Superfund site in 2001 and $100 million has been spent since, dredging PCBs from the muck and disposing of them. That's as it

* EPA actuaries calculate the price of a life at $6.9 million, which means that if it costs more than $7 million *per life* to clean up a watershed, then at present, we can't afford it.

should be. Smart regulation ensures swift action.[4] The richer our culture is, the faster it can clean up its mess.

But blanket regulation—like the hundred-foot buffers now being forced on nearly every single creek system in North America, regardless of need—destroys the economic health that is a prerequisite for cleaning up hazardous waste. On a nearly pristine island with severe restrictions on all development, after spending many hundreds of thousands of dollars on hypergreen development and building, I must now pay to plant a tree or a parsley patch. This is plainly irrational and, frankly, backward.

So I find myself tucked up in a coffee shop in Washington state, listening to men who spent their lives cleaning up hundreds of square miles of major hazardous waste sites in the United States, the Middle East, Asia, Canada, and Mexico, using precise metrics, without which their work would fail, people would die, and they would be ruined. They methodically eviscerate conservation science, its lack of fieldwork, its unproven assumptions, the absence of hard science, and the dearth of testing.

Buffer science arrived in the Pacific Northwest from the East in the 1990s. Buffers had been used with some success in the East, where logging, farming, and industrial production have been typically done near water. In some places, using buffers, a two-hundred-year history of damage had been reversed. But in the West, because it is messy and expensive to log in steep ravines, buffers were often left around creeks to begin with. It's hard to argue, therefore, that industrial pollution was damaging water in the West in any systemic way.

But if you control water, pointed out my coffee shop buddies, you control everything.

Don Flora,[5] who spent his career ("I guess I'm what you'd call an old-school scientist") with the U.S. Forest Service, explained how it happened.

"People began, for the first time, to consider the role of ephemeral or seasonal streams—they may or may not have salmon or fish in them. Some of them, even though seasonal, do support spawning fish.[6] So bingo, if we do it for fish-bearing streams, let's do it for the rest, too. What they didn't realize at the time was how many of these ephemeral streams there are up in the mountains of western Washington and Oregon."

"Buffers were imposed in '93 and '4," Flora continued, "and since then no one has gone back to see whether the buffers are the right size. The efficacy of those buffers is entirely unknown. Not a dime has been available to do the studies. Why? Researchers know they're too wide—it's not rational. But the people who run the shop in Congress, they like no-touch, no-cut—'We're saving the forest and the old growth,' and indeed they are."

The result was economic collapse. Almost 90 percent of the forests are now in buffers. Logging was halted; Forest Service employment crashed. Harvests were down to 10 percent.[7] "Maybe that's a good thing in its own right. But they're using buffers as a mechanism."

In fact, says Flora, buffers have limited utility. Some fish-bearing creeks might need a two-hundred-foot buffer; most do not. Wetlands are nature's filter; their *job* is to remove impurities and clean them out of the water trickling through rock seams into aquifers hundreds or thousands of feet deep. Up in the forest, where PCBs were never used near creeks, the restriction is absurd. Says Flora: "Coastal forests are clogged and rotting, tree immune systems weak, the trees malnourished. The net effect on water is to diminish its supply because the trees are so desperate for nourishment."

I trot this information home and feed it into an increasingly volatile mix.

CHAPTER THIRTEEN

Leviathan Cloaked in Darkness

Under the present conditions of state organization and national sovereignty, the life and liberty and property and happiness of the common man throughout the world are at the absolute mercy of a few persons whom he has never seen, involved in complicated quarrels that he has never heard of.
—Gilbert Murray, *The League of Nations and the Democratic Idea*, 1918

The ruined landscapes, destroyed communities, and shortened lives caused by the movement are ignored, because all members believe they are doing God's work. Or rather Gaia's work. As Alston Chase pointed out, ecosystem theory had enormous political potential, which had been recognized by Oxbridge authoritarian dreamers in the 1920s. Ecosystem collapse could be harnessed as the ideal first principle in creating a new kind of man, a better man, one less bound to getting and spending, more happily bound to the earth and its rhythms.

Rio, the UN's massively successful Earth Summit in 1992, was designed to create that better man, to reorder the use of land and resources with the aim of ending social ills: racism, sexism, classism, and, especially, ecosystem decline.

But to understand Rio's genesis, we have to go back to the

1976 UN Conference on Human Settlements, held in Vancouver, Canada. This conference, cloaked as it was in high sentiment, was the first formal incursion the United Nations made into national sovereignties through land use.

"Land," it was declared at that conference, " . . . cannot be treated as an ordinary asset, controlled by individuals and subject to the pressures and inefficiencies of the market. Private land ownership is also a principal instrument of accumulation and concentration of wealth and therefore contributes to social injustice."[1] Well, there you go: problem and solution.

For the United States, the signers were Carla Anderson Hills, who would become an architect of NAFTA and the World Trade Organization (WTO), and William Reilly, who would become one of the most effective heads of the EPA and subsequently chair of the President's Council on Environmental Quality under Bill Clinton. Neither were private-property freaks. In fact, they recommended abolishing private ownership of land piecemeal by restricting that ownership to sharply limited life leases.

Here are Hills and Reilly's most salient recommendations:

1. Redistribute population in accord with resources.
2. Government must control the use of land to achieve equitable distribution of resources.
3. Control land use through zoning and land-use planning.
4. Excessive profits from land use must be recaptured by government.
5. Public ownership of land should be used to institute urban and rural land reform.
6. Owner rights should be separated from development rights.

In rural North America, that agenda is now pretty much entirely in place. These and fifty-nine other specific recommendations would eliminate the cause—private property—of wars and racism, heralding a new age of social justice. Making it happen would require an overarching set of ideas that would hurry people toward understanding how wonderful it would be for all of us.

None of this stirred interest in the general public. In fact, no UN conference ever stirs much interest, other than among the conferees themselves and the requisite puffery from the world's press. The press's curiously supine respect for all things United Nations finds its beginnings in the work of Seán MacBride, a prominent international politician, public money trough–snuffler, and former head of the IRA.* In 1980, MacBride issued a statement called "Many Voices, One World," which proposed a "new world information order," and issued a report called *Many Voices, One World*.[2] This "new" way of reporting or communicating information would be coordinated through UNESCO.† One agency, the Institute for Global Communications (IGC), would monitor all information flowing from the UN's many conferences. Its various closed‡ computer networks (including AntiRacismNet, PeaceNet, WomensNet, and EcoNet) would allow participants and fellow travelers to communicate among themselves, wordsmithing their communiqués to the outside world into sweet pabulum. Reporters and reporting are discouraged. No harsh searchlight is allowed to shine into the magic.

And we buy the United Nations hook, line, and sinker. When its satellite, the International Union for the Conservation of Nature (IUCN), issued a report in the summer of 2010, the Year of Biodiversity, claiming that fully one-third of the earth's species are at risk of going extinct, ten thousand almost identical stories were published in almost every respectable newspaper in the

* Irish Republican Army.

† United Nations Educational, Scientific and Cultural Organization.

‡ Per privacy statement at www.igc.org/html/host.html, "closed" means subscriber-based Web hosting without automatic access by government agencies to customer records without a search warrant, which is convenient for plotting revolution from within.

world. Canada's oldest national newspaper, the *Globe and Mail*, sent its distinguished former *Comment* editor to the conference, and he wrote a scare story virtually indistinguishable from the others. When you can get one of the sharpest knives in the block parroting the party line, you have accomplished an impressive control of information. Not one story questioned one sentence of the conclusions sent down.

In 1945, evolutionary biologist Julian Huxley, brother of the famous fantasist Aldous, helped found UNESCO and went on to help found the IUCN in 1948. Huxley also cofounded the World Wildlife Fund in 1961, which in turn produced, in 1982, the World Resources Institute. Public policy recommendations flow out of these outfits like rivers of living water. Embedded in their high-minded aims are the origins of Rio.

The IUCN, the UN Environment Program (UNEP), and the UN Development Program (UNDP) consist of delegates appointed by member nations. In the United States, delegates are appointed from the departments of State, Agriculture, Interior, and Commerce, the EPA, Fish and Wildlife, the Forest Service, and the National Parks Service. Together they pay $519,000 in annual dues to the IUCN, as do other senior government agencies from 167 other countries. The IUCN is flush with cash and now spends, as detailed earlier, about $150 million a year promoting fears of biodiversity collapse. The IUCN also includes 875 NGOs, which work with the government agency members. NGO members include the World Wildlife Fund, the Nature Conservancy, the Sierra Club, the Audubon Society, the National Wildlife Federation, et cetera. All contribute money as well as expertise, and they work together to make policy.

IUCN directives are effected in rural America by the employees of federal agencies.

In 1987, Oxford University Press published a policy manual called *Our Common Future*, which arose from the World Commission on Environment and Development. In 1990, the IUCN, the World Bank, and three ENGOs published another policy manual, called *Conserving the World's Biological Diversity*, 193 pages of worthy sentiments about conserving nature, which caused the world's press and peoples to rejoice that someone was finally stepping up to the critical task. In 1991, *Caring for the Earth* arrived, another boilerplate high-minded piece of rubbish. These were just softening the ground, bombing before the storm troopers arrived. They arrived in 1992, when *Global Biodiversity Strategy* was delivered, 244 pages of detailed instructions on how to save the earth based on ecosystem theory. The strategy became the basis for the 1992 UN Conference on Environment and Development, aka the Earth Summit, or Rio. This was the conference that got the world on board.

The organizational structure for this environmental behemoth was developed by the titanic brain of Maurice Strong. Strong, who famously said he uses capitalist tools to achieve socialist ends,[3] is the chief architect of the UN's various environmental programs, the focus of which is the reconfiguration of land use to achieve social justice. Strong has played on the righteous desire for a healthy environment to prosecute his agenda for decades. One of his many roles was serving as the power behind Kofi Annan's throne, up to and during the Oil-for-Food program scandal. During that time, too, Strong shoehorned himself into nearly every Western government as a senior adviser. He is now tucked up in China working his magic, in part because questions remain about his participation in the Oil-for-Food scandal, but mostly because command-and-control "democracy" is definitely Strong's government of choice. In 2008, he wrote in *World Policy Review* that getting rid of the ballot box was the only thing that would resolve catastrophic climate change and the wholesale collapse of biodiversity.

So the participants in Rio produced, after a great deal of churning, dancing in five-star resorts, and swanning importantly through the hot spots of Latino culture, three very important documents. The first was the UN Framework Convention on Climate Change, which was the bed from which global-warming fever rose. The second was the Convention on Biological Diversity, which established authority and guidelines for land use. And the third? Why, Agenda 21, to which were attached 150 recommendations that would transform public policy to conform to the UN framework. Agenda 21 is a soft policy, not mandated, and it sounded so good, so filled with the right language—equity, social justice, peace—that no senior government official could afford to look as if he was against such sterling goals. Bush the elder did not sign the Convention on Biological Diversity, because it was binding, but he signed the UN Framework on Climate Change, because it was not. And he signed Agenda 21.

Clinton signed the Convention on Biological Diversity in his second month in office, and Strong surged happily onward in his drive for social justice. Conferences on Environment and Development, Human Rights, Population and Development, and Women followed, one a year between 1992 and 1995; it was boom time for bureaucrats with internationalist leanings. Strong promised that conferees' policy recommendations would alleviate "unsustainable consumption, social inequities, and poverty." Not unsurprisingly, population-control measures were considered critical to this goal.

Nineteen ninety-four's Earth Charter, developed by Strong and Mikhail Gorbachev, was a statement of ethical principles, which Strong said he hoped would become the new Ten Commandments. During the first UN Conference on Biodiversity,* a sixteen-page document, written in exhortative cereal box prose, mentioned that the 328 biosphere reserves in the world would

* Correctly known as the 1995 Conference of the Convention on Biological Diversity.

form an ideal core for a series of protected areas. This would be the first step in implementing the convention's guiding principles.

So voilà, here we are at 1995. Conservation biology has finally not only trumped old-school biology, it has erased it. Clinton, bowing to what must have been inexorable pressure from the environmental lobby, set up the President's Council on Sustainable Development, which produced recommendations on how to transform America into a sustainable country. The IUCN had decided that the best way to institute Agenda 21 was for each nation to build its own sustainable program, which the IUCN would then bring together under one treaty.[4] Agenda 21 must have seemed harmless to the 179 countries that signed on. Little would be lost and possibly much gained. Finally, in 1996, the *Global Biodiversity Assessment*, a 1,152-page blueprint, was published, outlining the road map toward a sustainable planet. It was in the *Global Biodiversity Assessment* that the first mention was made of the need to reduce population to one billion citizens who consume like present-day Americans or five to seven billion who would live much like the early settlers.[5]

The architects of Agenda 21 weren't quite so sanguine as to believe the agenda would roll into every township and county without resistance. So in 1995, the IUCN and its thousand satellites declared that they had pioneered a new kind of decision making. Rather than the confrontational, argumentative framework that marked the last 2,500 years of human progress, "consensus" would be a new, collaborative decision-making process. Designed in part by the RAND Corporation—their version is called the Delphi technique—for use in nation building, consensus building has been increasingly used in many townships, towns, and counties since 1995. I have sat in on a couple of weekends of "visioning" the future on my islands, which sent me into a tailspin of worry and fear. This is how it works.

A call goes out to all citizens and interested groups that a visioning process is about to begin. On the Pacific Coast, the Tides Foundation often stickhandles the process. Tides[6] funds a local group, which calls the conference. A prayer is said, usually by a member of the indigenous tribe in the area, reminding participants of their great work on this weekend or evening, the mission of saving the planet. Which, by the way, was much better tended when the tribes were in charge. Breakout sessions follow: people who are predetermined as contentious are put in groups with four or five true believers, who will work assiduously at "establishing consensus" by suppressing inconvenient thinking. Leaders gently guide participants toward deciding that all things green are at terrible risk and that something must be done. Leaders convene behind the scenes to track progress and identify the "good" participants and the recalcitrant ones. New leaders are assigned to those workshops that interest the nonconformists. Finally the assembled are read an unholy bunch of mealymouthed platitudinous crapola and told it is "their" vision for the future. Congratulations and good night.

It is fun to list what's wrong with all this extra-sovereign organizing, subverting of the press, and consensus building used to support the idea that "people" in general demand a green regime. But most salient, perhaps, is the criticism put forward by law professor and Hoover Institution scholar Kenneth Anderson, who in a recent essay pointed out the utter lack of legal liability at the United Nations and other international organizations. "It was not just the Oil-for-Food scandal, however, as those familiar even superficially with the opportunities for fraud, self-dealing, and rent-seeking in a system at once as byzantine and unaccountable as the UN's would recognize. As more rocks began to be overturned in other UN programs and organs, evidence of serious graft, embezzlement, kickbacks, and other financial fraud of a kind that would plainly be criminally prosecutable, if only there were someplace to prosecute it, began to emerge in other UN programs."[7]

Nevertheless, bureaucrats at the United Nations, in concert with federal agencies and ENGOs, set the agenda for land-use planning worldwide.

Because no voter elects anyone to the United Nations, no voter can hold it or any of its army of functionaries to account for their actions and fire their sorry asses. Nor does anyone within the UN seem to have any idea that its policies lay waste to the natural world. In fact, the only local government able to prosecute a UN official is the attorney general of New York State or the New York district attorney. Which is preposterous.

Anderson concludes that the United Nations has massive structural failings. No holding of any individual or institution or ENGO to account means that collapsed rural economies and degrading forests and ranges—the results of thirty years of IUCN activity—only exist in the real world, not in the world of the UN.

The final piece of the puzzle snaps into place when you look at the activities of various corporations operating within the beast. In many rural areas, draconian regulation is supported by the multinationals and the larger corporations.

They support the regulation for three solid reasons, the first being that it's simply good business to co-opt your enemy—in their case, ENGOs. The second reason is also good business. Ecosystem management drives small competitors out of business. Finally, if you are certified "green," you can charge higher prices! Therefore, it would be bad business *not* to get in bed with green.

In Canada's boreal forest, an area twice the size of Germany, multinational forestry firms work with multinational ENGOs to determine the use of Canada's natural resources. Local operators whistle in the wind or beg for a green job. In the Alberta oil sands, North America's windfall cache of oil, the Pew Foundation, founded on the oil sands' prime move, Sun Oil's profits, lobbies ceaselessly for regulations that end by constricting smaller operators while leaving the larger ones, including Sunoco, more powerful.[8]

In nearly every country in the world, big ENGOs broker deals between national and state governments starved for cash and multinationals hungry for profit[9] for the use of that country's natural resources. The national government receives a hefty payout, and the ENGO trumpets a save, but can citizens develop their own industries? That ship has long sailed. The taxpayer picks up incidental costs, unwittingly burnishing the multinationals' balance sheet.

Some mainstream columnists (Jonathan Adler, Margaret Wente, George Will) have asserted that, with the virtual collapse of the climate-change argument, the green movement is dead. Hardly. It's gained a massive new ally and is running ahead of the wind.

The Texas Department of Transportation (TXDOT, pronounced TEX-dot) is the richest, most powerful state agency in the United States. Overseen by a crew of big swinging dicks, it builds roads, loves building roads, and has never seen a road it did not want to build. For almost a decade, deliberately neglecting all other roads in its service, TXDOT's big enchilada was the Trans-Texas Corridor (TTC), the first leg of the mythic NAFTA superhighway.

The NAFTA superhighway was conceived to transport goods from a Mexican port city built by the Chinese to big-box stores throughout North America, cheaply, easily, and fast. Four football fields wide, it would be a "multimodal" transportation corridor 1,800 miles long and 1,200 feet wide, costing about a trillion dollars. It would include tollways for passenger vehicles and trucks; lanes for commercial and freight trucks; tracks for commuter rail and high-speed freight rail service; depots for all rail lines; pipelines for oil, water, and natural gas; and electrical towers and cabling for communication and telephone lines. And on either side, two vast strip malls.

Opposition to the NAFTA superhighway from people on both sides of the political divide was so intense that to date, states all along its putative route have passed resolutions against it. The ideal spur to local organizing, even the rumor of it inflames public debate.

And now it is supposedly dead. In fact, it was a myth, snickered Clay Risen in the *New Republic* in 2008, a paranoid vision from the paleowacko contingent. In its essentials, however, the NAFTA superhighway plan exists today under different names, the West-wide Energy Corridor and the Asia-Pacific Gateway and Corridor initiatives among them. If those prove unpopular, the names will be changed again.

The first leg of the NAFTA superhighway would have been the Trans-Texas Corridor. Plans had been advancing steadily since 2000, to the point where Cintra, the Spanish transportation giant, had contracted, by paying a billion dollars up front, to build and operate the first 550 miles of the behemoth in a public-private partnership with the Texas government, which had joyously passed the legislation in 2003. Cintra is owned by the Spanish royal family, who would own the highway, with every business along it paying rent to that family. Aside from the initial payout and business taxes, the state of Texas would have had no revenue from the tolls. The goods coming in from Mexico would have been inspected first in Kansas City, Missouri, by Mexican customs officials, in a 250-acre complex already built. Those who say the NAFTA highway was a populist scare story claim the complex doesn't exist, but many people have seen and photographed it.

Fred Kelly Grant explains the setup.

"China had developed and is operating the ports of Manzanillo and [Ciudad] Lázaro Cárdenas, and the TTC was going to lead directly from those port cities through the interior of the United States, meaning that no goods from China would ever be inspected by U.S. Customs," says Grant. "I couldn't understand why the Chinese would go to that trouble when they already own the Port of Long Beach, which they bought during the Nixon administration. Some union people explained it to me. First of all, with a port in Mexico, there would be no Teamsters Union involved, which means that expense is gone. No union longshoremen to pay, no truckers. And, when they ship something into the Port of Long Beach, there's no U.S. Customs inspection, but when they truck it out of there, there's a chance someone might want to inspect those trucks. By building the port in Mexico leading to a noninspected freeway, the Chinese would avoid U.S. Customs all the way up the line."

For this, TXDOT chief Rick Williamson was more than willing to stand up to the mounting opposition. Called by *Texas Monthly* "the most hated person in Texas" (also called Darth Vader by a member of the Texas legislature), Williamson told a reporter from the *Nation* in 2007, "We're the greatest state agency you'll ever interview," a hubristic statement that places one in a kind of fascinated awe, hoping to be around for the fall.

Williamson's flank was the adamantine wall against which hundreds of thousands of Texans subsequently hurled themselves. All the usual tactics were used: tens of thousands marched on Austin. Then again. And once more. There were uncounted letter-writing campaigns, phone-tree manipulations, presentations by every interest group to any member of the legislature who would sit and listen, fund-raising concerts at county fairs, and the vocal shifting of public allegiance to politicians like Senator

Kay Bailey Hutchison (who promised to stand against the TTC). Depositions were made and opposition bills were introduced: all the tools of the powerless in the face of bureaucratic will. Media noise accompanied every sally. Advisory panels of citizens who lived alongside the proposed route got involved. The Farm Bureau pleaded for the project's death. And at the start, there were masterful campaigns by the usual environmental groups inveighing against the loss of habitat.

But the TTC lobby was spending $6 million a year for parties, gifts, and posh travel for legislators and their aides, and with all that high-powered churning, it was a while before anyone noticed that the big ENGOs had switched their support and negotiated the sequestration of three acres in other parts of Texas for every acre taken by the corridor. It turned out that the Nature Conservancy and the Sierra Club were in favor of a monstrous strip mall cutting through the heart of America.

For Texas, the corridor meant ten-lane highways, six rail tracks, utilities, pipelines, state concessions—gas stations, restaurants, motels, stores, and warehouses—on four thousand miles of toll roads that would consume more than eight hundred thousand acres of Texas taken in a combination of outright purchase, condemnation, and eminent domain—a looting, therefore, in a state where 94 percent of land is private property. Breaking that kind of hold on private property is a dominant goal of Agenda 21.

With the loss of the lobbying power of the environmental movement, all protest appeared for naught. There was simply too much money to be made.

Then one day Margaret Byfield called up Fred Kelly Grant. Byfield is a legend in property-rights circles. Young, tall, and slim, with an impressively cool head, she is the daughter of Wayne

Hage, the Nevada rancher who came to typify the Sagebrush Rebellion.*

It is impossible to understand what happened next in the Texas corridor without learning about the Hage lawsuit. The Hage case flushed out all the players and their quiet maneuverings, laid out the collaborations between ENGOs and federal bureaucrats, and revealed the gauntlet of conflicting regulations seemingly devised only to strip rural dwellers of self-determination.

Hage and his family fought the federal government in court for eighteen years over the taking of Hage's ranch's grazing allotment in Nevada, his water rights, and even his cattle. The Hages won every single battle, and finally, in 2008, after the deaths of Hage's first wife, of Hage himself, and then of his second wife, Congresswoman Helen Chenoweth-Hage, the Hage estate was awarded $12.8 million in the case against the Forest Service. Hage had trained as a chemist, and his mind was suited to the excruciating detail required to deconstruct agency onslaughts.

Hage's ranch and its water and grazing rights had been worked since 1860. Hage had bought it in 1978 for $2 million. It was high-desert ranching; cattle need a lot of land up there at the top of the world, so the deeded ranch was seven thousand acres, and the public-land section was eleven square miles. Life for the Hage family was good—five golden kids and two thousand head of cattle. For the Hage family, the care of the land had always been a cherished responsibility, and they worked it the old-fashioned way, on horseback. Like every other working ranch family, they had vested grazing and water rights on their leased federal lands, which were taxed as Hage's private property.

In the 1970s, parts of the Forest Service and the Bu-

* Byfield maintains that her father was not part of the Sagebrush Rebellion; he did not think that public lands should be sold to private citizens, only that leased lands had property rights vested in those leases, that they were taxed accordingly, and that the leases could not be overturned.

reau of Land Management decided to reinterpret federal land management—as well as property law and water rights. Politicians fanned out to look at those lands and determine their future use. Would water be taken from the ranges and sent to the cities? Or would the land be reclassified as wilderness or a wildlife refuge? Every desirable piece of land leased by a rancher was considered, and plans were made to repurpose its use. Hage heard that Harry Reid and his wealthy backers had camped up on Table Mountain, on his allotment, and had decided that this place, where cattle now ran, would make an ideal elk preserve.

Out rounding up cattle one day in 1980, Hage and his crew saw coming down the trail several men from the Forest Service. Hage did not recognize the usual faces; these men were strangers and said they were from the Austin office. The agents explained that the Forest Service was filing a claim on all the water in the Monitor Valley. Hage thought that his water rights were vested and part of the ownership of his ranch. Why would the Forest Service be filing on his water rights?

"Because," the agent responded, "that is what we were ordered to do."

Hage contacted Nevada's state engineer, who confirmed that the Forest Service and BLM had filed claims on 160 vested water rights held by Hage and the Pine Creek Ranch. Hage's only recourse was to petition the state engineer, requesting a determination of who had what rights in the Monitor Valley. This petition was filed on October 15, 1981. Adjudication, which should have taken months, stretched out to ten years.

After years of waiting, fee paying, and legal costs, his water and grazing reduced every year, Hage found himself on the brink of financial collapse.

The delays were deliberate. According to the Hage family, documents obtained under discovery during the trial show that

District Forest Ranger David Grider sent a copy of the cancellation notice to the attorney for the National Wildlife Federation, Roy Elcker, and in that same letter thanked Elcker for NWF's lobbying efforts in Congress on behalf of an increased Forest Service budget. NWF's policy has always been aimed at ending all grazing and agricultural water use on "public land." During a lecture before other environmentalists, Elcker declared, "How you win is one at a time, [the rancher] goes out of business, he dies, you wait him out—but you win." The Forest Service and the environmentalists were past masters, continued Elcker, in the art of "making it so expensive to operate, and make so many changes for him . . . [that] he goes broke."[10]

Agents began packing guns on the Hage allotment. After Hage had been informed that his allotment was closed, on three occasions heavily armed agents prevented Hage's employees from moving cattle off the closed allotment. During these intrusions, 104 cattle were confiscated and subsequently sold at auction, the profits remaining with the Forest Service. Agents sent the family a bill for the costs of confiscating the cattle.

Most ranchers would have given up, overwhelmed by the complexity of the fight and the cost of battling the agencies in court. Instead, Hage filed suit.

The Sierra Club, the National Wildlife Federation, and the Natural Resources Defense Council immediately filed for status as "intervenors," arguing that ranchers should receive no compensation for losing their water and grazing permits on federal land. They were joined by the Nevada attorney general, who claimed that ranchers had no rights on their allotments. But the U.S. Court of Federal Claims denied the environmental groups' motions to intervene. And during the eighteen years of motions, trials, appeals, and countersuits, the tangled web of government connections with the powerful environmental movement was laid out in document after document. The Hage family watched as every federal and state agency wrote more regulations every year. The whole family learned to deconstruct each new set of regulations at the dining room table, and more important, learned how to bend the complexity to their purpose.

Margaret Hage was ten years old when her father started fighting, and after college she started Stewards of the Range, a foundation to help her family fight the government. Today, at forty-three, she travels ceaselessly, teaching rural Americans how to "coordinate" to save their towns and economic lives. "Coordination can rebuild rural America," she says. "It's just that powerful."

Margaret had met Dan Byfield, a water lawyer, on a committee that was trying to create an alternative to the Endangered Species Act. "He was the only one who didn't sell out," she said. "He was a keeper." Dan moved her down to Texas as soon as he could. The couple have a daughter, now seven, and have built a house on a piece of flat land in Bell County, in the Blackland Prairie district, north of Austin. Dan built Margaret a one-mile running track through a forty-acre grove of live-oak and pecan trees, which bordered on the Little River. For the first time in her life, she had a home where she felt safe.

Then she discovered that the Trans-Texas Corridor was going to be built a mile from her house. "I'm not a curse," she said to her old friend, Fred Kelly Grant.

"Let's stop this," said Grant. "I know we can." That's what he said, though privately, he thought they were toast.

Another rancher stepped up. Ralph Snyder, who stands about six five, and his wife, Marsha, who is about four nine, had ranched the region for decades while raising five kids and running a profitable parts and salvage business. He had never—as ranchers tend to say—had a real job, never punched a time clock. Therefore he was available for organizing duty. Within weeks, he had put together a coalition of four small-town mayors and the commissioners of four school boards. Not one of the towns had more than a thousand residents.

The four mayors informed TXDOT that they were ready to coordinate.

The TXDOT people burst out laughing. Then, once sober, they called their lawyers, who informed them that they had a duty to coordinate. In fact, it was the law.

During the Hage family's two-decade fight, Margaret had learned to concentrate on one thing above all else: preparation. She and Dan put together a citizens group to do research. The group consisted of Marsha Snyder and her best friend, who both worked fourteen-hour days for twenty-seven months. TXDOT had thirty days to respond to the request for coordination. On day twenty-eight, the agency set up a meeting.

"We knew what they were doing for those twenty-eight days. They had all their lawyers trying to figure out how to get around us," laughs Margaret.

"Preparation was key," she repeats. "In our first meeting with TXDOT, we discovered that TXDOT had not done the assessment right. We had looked at the law, looked at the regulations behind the law, read the environmental impact statements, looking for where the agency work was insufficient. And so they were. Their environmental study had no details. Five hundred and fifty miles, which would take 146 acres per mile, and they said the only economic impact would be a change in land values.

"There was nothing about school districts, the loss of revenue from farmlands, the loss of productive use, the loss of revenue for towns. They passed on every decision, referring it all to the Tier 2 study. Tiering regulations say that Tier 1 is broad based and an overview but has to be detailed as well. You can't ignore impacts the way they had. Ralph had the tiering regulations in front of them, and he began to ask them what they thought they were doing.

"It was a deer-in-the-headlight moment. There were seven TXDOT people across the table, including [Ed] Pensock, who was the director of corridor systems. He's fun to watch. When he gets rattled, when his people get out of line, he just talks nonsense over everyone.

"We established we had done our homework. It was a four-hour meeting; we did not let them take a break. We like to imagine the atmosphere in the car on the way back to Austin," she laughs.

Margaret then sent a letter to the EPA informing them they were ready to coordinate. The EPA thanked them for their "comments."

"We sent the EPA a letter saying we were not commenting. We were requiring them to coordinate the environmental study with us. We said we had concerns that were not being addressed."

The EPA did not respond.

Margaret then wrote the Department of Justice and copied the EPA chief in Region 6. Representatives of both agencies were on the phone to her within hours of receipt of her letter, offering possible meeting dates.

"We then wrote TXDOT, saying we really needed their full Tier 1 analysis for our meeting with the EPA, and within minutes they called to say they were going to come out and have that second meeting a couple days after the EPA meeting. We told them, 'You've refused to set meetings with us two times, and we need that info before the EPA meeting.'

"The EPA meeting demanded a different preparation. We focused on the deficiencies in the environmental impact statement, because that is what the EPA oversees. Our school superintendents explained that the TTC would put on a hundred miles in school-bus routes every day; because the exit ramps were ten miles apart, the highway would sever our counties. All the fire districts are serviced by volunteers, who raise money with bake sales. They would have to raise money to build new fire stations, because they wouldn't be able to reach some houses in the time it takes to save a life.

"We would have to redistrict school districts, a huge expense, which wasn't considered. We all used the farm-to-market roads. They would be cut off."

"It was so blatant," says Ralph Snyder, "it was unbelievable. Cintra would take the tolls, sell pad sites, get all that money. Cintra would charge us for the water and electricity lines that would cross this thing. We would have to pay to dig up all the lines, which now lie fifteen feet deep, encase the lines, and lower them another ten feet, and then they'd charge us rent to cross the highway. There were nineteen water companies in seven counties on the east side who had to cross the highway. The golden commodity in this state, water, would be even more expensive. Power lines? Same thing. If you wanted an overpass, the taxpayers had to pay for it. And Cintra could put a toll on it. And if you broke down on their toll road, the towing company had to pay Cintra for the right to come fix your car. All the normal rules on public highways didn't apply. Since it would cost you to get off the toll road, all those businesses who now depend on highway traffic would be bankrupted.

"These were completely new issues for the NEPA [National Environmental Policy Act] reviewers and the NEPA attorney. They had no idea how rural America worked. They were taken aback.

"We made sure all our people talked first, the four mayors, the fire chief, schools superintendents. Then we turned it over to the EPA. That was so we didn't ask a question and they would spend thirty minutes with a bunch of gibberish, drawing out the clock. We got all our stuff on the record first. We put together a manual for the EPA, with all the things that TXDOT was supposed to look at but didn't, including the transcript of our first meeting with TXDOT."

Mae Smith, mayor of Holland, a firecracker of a woman, laughs today. "First time, there were two of them, they came up. They thought it would be fifteen minutes.

"At the next meeting with TXDOT, they brought a transcriber, lawyers—twenty people showed up. We started throwing things at them, like the Farmland Protection Act, the State Wildlife Action Plan, and then we brought up the Blackland Prairie."

The Blackland Prairie is flat, easy to build in, part of a black-dirt swath that runs up the center of the country; in some places, it is hundreds of miles wide. In Bell County, Texas, during the Depression, more people farmed here per square mile than in any other part of the country, because it is the most productive dirt in the country. When I visited, it had not rained for eight months, but still the corn and sorghum were green in the fields—curling, but living.

"The Texas Blackland Prairie is like black gumbo, the most fertile land in the nation, and they were going to run right through it. Why didn't they mention that in their report? we asked.

"We met with Texas Parks and Wildlife and found that they had told TXDOT *not* to go through the Blackland Prairie. One of the great advantages is that these agencies work together. They all keep records. They say one thing there and another thing here. You can really embarrass some people.

"We piled on the research and delayed the publication of their environmental study for eighteen months by asking question after question, exposing one inconsistency after another. It's not magic.

You study the law; 99 percent of the time, they are not following the rules. You make a list of things they haven't done."

"We found fifteen separate laws that they violated," says Fred Kelly Grant.

It took five small-town mayors, five school superintendents, Ralph and Marsha Snyder, Dan and Margaret Byfield, Fred Kelly Grant, and two ranching wives who performed the research twenty-seven months to shut down the biggest infrastructure project in recent American history. They defeated the most powerful state agency in America and a multinational so hungry for profit and so confident and habituated to running over local wishes that it didn't bother to do its homework. That's how powerful coordination is.

On my last day in Bell County, Ralph and Marsha take me to Bible study at their country church (discussing Revelation 4 that day) and then to church itself. It is Mother's Day, and the new fathers in the congregation smilingly pass out flowers to every woman in church. There is a lot of laughing in both class and congregation, a lot of the sort of hugging and sweetness that has almost vanished from our communal lives. Afterward we go to lunch at Flag Hall, the remnant of a Czech lodge founded by Bell County pioneers a hundred fifty years ago. Five hundred people eat Texas barbecue, cooked out back in fire pits by county men, and potato salad, beans, bread, coleslaw, and a cornucopia of desserts made by county women. There are four generations present in the hall. It is loud, and gusts of hilarity rush through the room. Bell County, despite the drought, despite the economy, stood strong and fought off the encroachment of the superstate, which has ruined so many other county economies. They, at least, have reason to be happy.

Victory

Apparently mockery and derision unbalance conservation bureaucrats, habituated as they are to forelock-tugging deference. Sensing, correctly, that the next election is not going to go their way, they decide to ram through buffer regulation at double speed. We force rewrites by pointing out the bad science, the lack of mapping, and the fungible metrics. After three hasty rewrites, the bylaw becomes incomprehensible, perfect for our purposes.

By May, when the public hearing and voting are scheduled, the whole island is in a happy uproar. The weekend before the public hearing, we plant 250 signs in every major runoff ditch system. The signs proclaim that fifteen-to-hundred-foot setbacks mean no gardening, no mowing, no pruning, no touching on more than twenty-four thousand acres of the island. Within a day, the signs are gone. We replace them, and that afternoon, Jamie, hearing an engine outside our gate, throws it open and sees a red pickup truck, with our signs in the bed, careening around the corner. The driver sees him, leans out the window, shouts something, and shakes his fist.

The public hearing is thronged with too many people and has to be rescheduled.

The next night, six hundred people—20 percent of the resident island adults—turn out. As does the RCMP, who we suspect were called by the trustees in a ploy for sympathy. They are armed, wearing flak jackets, their SUVs parked diagonally across the driveways, ready for action.

Elderly farmers still in their overalls sit in the back, wives beside them; few have ever attended a trust meeting. As people stream into the building, I joke that if they have muscle tone, they're on our side. We take unseemly delight in the discomfort of the trustees sitting on the stage and watching slack-faced as people keep arriving. Christine's square face is turning to stone. George is swallowing convulsively. The face of the off-island trustee, Sheila Malcolmson, who will cast the deciding vote, is flushed red and blotchy.

There's an overflow of 250 people, who are placed in a holding room with a tinny speaker, and the meeting is called to order. Twenty minutes of procedural droning is interrupted by a young woman who forces her way past the armed cop in a bulletproof vest and declares that the speaker is broken and that therefore the public hearing is void. To shouts of "Dump the trust" from the back of the room, the hearing is postponed again.

The next day, the bylaw is shelved. We are the only jurisdiction anywhere in North America that has turned back the hundred-foot buffer rule.

It may be temporary, but we have our first substantial win; it is one of the few in trust history, and a few of us marvel at how easy it was to stop them, once everyone got engaged, but how hard it was to get people off the couch, including ourselves.

"My generation abandoned their responsibilities and walked away," says Fred Kelly Grant. "We deserve what happened." It is a truism that the adults in many towns and counties, the traditional town fathers and mothers, ceded their government to professionals, whereupon the fanatics crept in on cat feet. "Taking the country back," in the maligned phrase of the Tea Party, means more than rousing town-hall meetings and rallies; it means plowing through tens of thousands of pages of bylaws, regulations, and rules that have constricted the lifeblood of the country and, while keeping the essential, discarding the rest. It is a paralyzingly tedious job but necessary. And watching people doing that, digging in, engaging with the levers of government, gaining stature by the moment, is a miraculous sight, as beautiful as Venus emerging from her clamshell.

For me, engaging in local politics, not as a complaining activist but as a problem solver, conferred an almost immediate increase in self-respect and, often enough, heady pleasure. I quit a half-dozen times, certain there were better ways to spend my time—plus, there is no truer truism than that politics makes strange bedfellows—but eventually I crept back, ashamed of myself, sat down, and did the work. Every highflier of my generation would find substantial character improvement and much-needed humility by the effort of effecting real change—actual, useful change—the achieving of which is harder and more brutal by far than rioting in the streets or climbing any career ladder in any profession now enthralling the popular mind.

The trustees realize they will not win, so decide not to run again. But Jill Treewater's videotapings reveal that they have funded several trust advocacy committees, which will operate as lobbying and rule-making bodies going forward. Both George and Christine either sit on or chair these committees. We sue them for conflict of interest, demanding they return the money, and that they are substantially fined.[1] And we go on to defeat their proxies in the subsequent election by two to one.

CHAPTER FOURTEEN

It Is All About the Water

Whiskey is for drinking; water is for fighting over.
—ATTRIBUTED TO MARK TWAIN

Chuck Leaf promises to meet me in the parking lot of the Texas Roadhouse in Greeley, Colorado. He is waiting in his light green Cadillac, so I knock on his window, and we run through sleet to the restaurant door. It is 4 P.M., suppertime in ranching country, the atmosphere boisterous, so we pile into a booth in the farthest corner of the place. Peanuts spill over from a tin bucket, the music is loud, our waiter so bubbly he's like a cream soda. We both order steak.

Dr. Leaf is an engineer, a hydrologist who started his career in 1962 with the U.S. Forest Service. He was in the research arm, a bench scientist. "We had a lot of field projects, and I was publishing a lot, but they were grooming me for something else. They'd send me to these weird meetings in Washington to teach me how to manage people and their minds—it was crazy. So I questioned them. Then they began to take our project money away and give it to universities for esoteric work which didn't mean anything, and they gave me the job to review all that stuff, give it credibility. I got to the point where I said, 'This is junk science. I'm not signing off on it.' That gave me immediate problems. My supervisor called me in and said, 'Chuck, we have big plans for you, but you have to adjust your attitude.' I knew my days were numbered."

Leaf's ranch lies on the South Platte River. It also lies in the Wildlands corridor[1] for Colorado, and his free use of his land has been reduced in increments by the state to 20 percent of what it once was. But he stays. Like many country people, he is wedded to his land. In the years when his water use was cut to nothing and his neighbors were driven out of business, he bought what he needed and soldiered on. He did well in private practice, but the work he began as a young scientist was too exciting to drop. He continued with it, financing it himself, testing it in the Fraser Experimental Forest again and again. In engineering, models *can* be tested authoritatively, and he methodically assembled his case.

The Platte rivers originate in the Colorado and Wyoming Rockies, join and drain the great basin of Nebraska, then flow into the Missouri. Here is what current Wikipedia thinking is on the Platte River:

> *Since the mid-20th century, this river has shrunk significantly. This reduction in size is attributed in part to its waters being used for irrigation, and to a much greater extent to the waters diverted and used by the growing population of Colorado, which has outstripped the ability of its groundwater to sustain them.*

Every word, including the *and*s, the *its*es, the *which*, the *by*, and the *in*, is wrong.

Leaf says that not only does eight million acre-feet of water—"a veritable ocean"—lie in the alluvial reservoirs under the South Platte (leaving farms and rural communities at perpetual risk of catastrophic flooding), but that simple, easy, cheap management of the upland forests could increase water flow by five hundred thousand acre-feet[2] every single year.

Even without tapping the dangerously supercharged alluvial zone on the South Platte, that's enough water for all the farms and communities and cities on Colorado's Front Range, enough for the farmlands of Nebraska, and enough for the endangered and threatened species on the Platte River system: the piping plover, whooping crane, pallid sturgeon, and interior least tern.

Colorado now sends ten thousand acre-feet a year to the Dakotas for the whooping crane on its migration stopover. The state has curtailed irrigated farming on the Front Range by 40 percent and is tapping groundwater aquifers for its new subdivisions. According to Leaf, those aquifers are dropping like boulders.

"It's like we farmers have a yellow star stitched onto our clothing," says Leaf, the heated language odd from such a determinedly ordinary man. "It's like in *Doctor Zhivago*, when members of the proletariat move into your house. We can use the water from our wells, but we have to replace every single cupful. Some farm wells have four meters on them. I forgot to read my meter for one day and received a reprimand from the environmental supervisor of the district within hours. And my water right goes back to 1866."

He pulls maps and photos from his briefcase, and the restaurant table becomes a desk. Leaf starts with the subalpine zone, the headwaters of the South and North Platte rivers. "It's lodgepole pine mostly—a lot of it is old growth. It's like a whole bunch of old people standing up there." He points out a photo of one of his experimental clearings, the edges jagged, like a horizontal rip in a piece of fabric; the tear is tiny relative to the expanse of forest.[3] "This is the way we used to take out hunks of old decadent forest and gradually replace it with younger trees, so you have age diversity. Now we have no age diversity.

"If we do this kind of cutting in the subalpine zone, the water increases persist for eighty years, and the increases are steady. As well, we found that if we do it in the high elevations, the montane zone, snow country, where there are spruce and ponderosa pine as well, it changes the aerodynamics of the canopy surface. We subject the watersheds to small patch cuts, protect the snow from wind. It means there is more snowmelt for the creeks and more water downstream."

Leaf went back into the historical record that measured stream flow, using the water gauge in Saratoga, Wyoming. "We have the hydrograph record since 1904. From 1904, when Teddy Roosevelt reserved the national forest and Gifford Pinchot started modern conservation of the forest, we lost 116,000 acre-feet of water each year in the South Platte and 135,000 acre-feet a year in the North Platte.

"There's not been much timber cut since the late '60s, '70s, so this"—he presents another slide—"will be the increment of decrease in the next forty-five years, if things are left to their own. Unfortunately it's not a static situation, because these old stands are being invaded by pine beetle and budworm in the North and South Platte and in the Colorado River basin. The end result is desertification, but first the runoff increases are very high, meaning floods. We begged the service to put these stands under management, because we knew if they didn't, the consequences would be what we are seeing today."

This is how absurd it is: there's not enough water for the farms that feed the urbanites on Colorado's Front Range, and the wheat fields of Nebraska are water-starved, but an ocean is supercharging under the South Platte and already flooding basements and farm fields. Groundwater aquifers are dropping fast, but there is a proven plan to increase downstream water flow. And conventional wisdom trumpets that there is not enough water.

Leaf thinks it's deliberate. "There's more than enough water for everybody and more."

• • •

A wetland is a holy thing, and wetlands have haunted the national discourse for decades. Say "protecting wetlands" and people's faces assemble into po-faced gravitas. What many don't realize is that wetlands are nature's state-of-the-art filters, its best mechanism for removing toxins and impurities, whether made by nature or man.

Wetlands, because of their importance in groundwater hydrology, have a distinct definition in law. Soil, plants, and the standing water itself have to demonstrate specific metrics before they are characterized as constituting a wetland. Geology is so important to wetland function that the senior, senior scientist in the game is a hydrogeologist. This is hard science, not conservation biology, not romping around in meadows quoting Latin names for common weeds, not cut-and-paste jobs by guns hired by the Pew Foundation, not "wetland specialists" who've taken a four-day course given by the U.S. Army Corps of Engineers and hung out their shingle.

What you need is someone like Steven Neugebauer,[4] who is a licensed hydrogeologist. Happily for my purpose, he lives outside Seattle, where the Department of Ecology, in response to the imminent polluting of Puget Sound,[5] instituted a storm-water program thirty years ago to divert storm water away from the sound and into groundwater sinks or wetlands, hoping that the pollutants would be filtered out that way. Results? Disastrous.

There's a lot of runoff in a city where torrential rains are legendary. But running floods of storm water into wetlands kills them, just as running storm water into drinking water reservoirs pollutes drinking water. Both actions? Illegal under the federal Clean Water and Safe Drinking Water acts, respectively. So Neugebauer, along with an associate, Justin Park, who is both a lawyer and a fluvial hydrogeologist, have launched a series of citizens' lawsuits against municipalities who are running their storm water into wetlands on both public and private property—which means pretty much all of them—and into drinking water and kettle lakes owned by the county. Fines start upon receipt of notice of intent to sue and run at $35,000 a day. So far, their target bureaucracies have run up $800 million in fines.

"Conservation biology's assumptions are all subjective. It's not a true science. We go into these things completely openly; whatever we find, we report. You don't like it? Tough. The other problem is that there is no standard. These so-called wetland specialists are not required to know hydrology; they're not required to know soils—all they look at is vegetation. The vegetation definition in the Corps of Engineers rulebook is biased; they include vegetation that cannot exist in a true wetland."

So the result of letting biologists with an agenda run amok? Polluted drinking water for the citizens of Seattle and its satellites and wetlands that have been flooded and killed.

One good thing: clogging the forests with trees decreases runoff, at least until the trees die because there are too many of them, their immune systems are compromised, and beetle kill finishes them off. Then the slopes above the gleaming cities of the West Coast degrade, and one year there is a flood of biblical proportions.

In 2001, farther down the coast and two hundred miles inland, the first shot in the water war of the twenty-first century was fired. Klamath Falls, once surrounded, like most rural towns, by an oasis of abundant farms, ranches, and forests, was about to have its water rights canceled. The Indian tribes in the area had made representations that the shortnose and Lost River sucker fish were critical to the tribes' spiritual, cultural, and traditional food uses, and that "additional development in the form of irrigation construction for the Klamath irrigation project altered

Klamath Lake and the surrounding water network that negatively impacted suckers and other fish populations."[6] Barbara Peterson, a retired Folsom State Prison guard, describes what happened.

"I've spoken to Indians who weren't in on the deal. No one eats the sucker fish anymore; it's disgusting; it's a baitfish. And ten years before they declared it endangered, it was such a pest, they tried to get rid of it by dumping poison in Klamath Lake.

"The federal and state agencies engineering the water taking called a meeting at the fairgrounds. They were going to explain to us what was about to happen, why all the farmers in the valley were going to be put out of business. There were snipers in full riot gear on top of the buildings in the fairground—black helmets, flak jackets, sniper rifles. For all us old people about to be ruined."

Peterson had just retired. "At Folsom, it was all yes sir, no sir, yes ma'am; things were clear. Here, nothing was what it seemed. Everything was a lie. I was standing in the shower one day, going through the list of lies,[7] and then I thought, *What if everything is a lie?*" That day Peterson became a rural activist.

Fourteen hundred farmers were affected, and hundreds more who supply the farms. "It's a dust bowl from here to Sacramento now," says Peterson. "I grew up in Sacramento. You used to be able to throw a seed in the ground, and it would grow without any help. Now, the land is desert."

Today, fifty miles south of Klamath Falls in Siskiyou County, all hell is breaking loose. Siskiyou is the test case of water taking writ large, a taking that California Fish and Game has told ranchers and farmers is being tested here, and once the right strategy is determined, will be moved across the country in a blanket cancellation of all community and agricultural rights to water.

The taking is based on the coho salmon and its supposedly endangered status in the region.

The coho is the nomad of the sea. It ranges all the way to Russia and south as far as San Francisco, but its core habitat is the cool seas of the north coast. The coho, unlike the king or chinook, summers in the shaded pools of the rivers of North America, and it is primarily the coho that forms the basis of all endangered-species salmon assertions that so terrify the general public. The coho is the reason for my salmon pond, three miles from the sea. Kathy's calculations are that one unheralded day, a coho will come up the river to spawn, jump the barrier between creek and pond, then summer in the creek, shaded by the willow trees. It hasn't happened yet. For one thing, the pond goes dry in the summer, and no one is absolutely certain that any salmon has ever been seen in my creek. Nevertheless, Kathy dug down and sank two deep concrete culverts upright in the pond, for shelter of last resort.

At issue are four dams, which have lined the Rogue River since the early 1900s. Owned and operated by Warren Buffett's PacifiCorp, the dams provide clean, green, local, emission-free hydroelectric power to seventy thousand homes and businesses in California and southern Oregon. But the dams are up for expensive relicensing, and the movement and agencies have convinced PacifiCorp to allow them to be removed, making a deal that PacifiCorp will be licensed to provide power—expensive dirty coal power from other states—to the remaining residents and businesses of the Siskiyou after the dams go out. The state will pay to remove the dams and indemnify PacifiCorp against any damage caused by dam removal. It's a win-win-win for Mr. Buffett, who appears to make money just by breathing in. Indeed, power rates have gone up 20 percent since dam removal looked likely.

Each dam has a holding reservoir, and the stored water is released to the ranches and farms downstream for irrigation. Over the past 150 years, these rights have been tightly negotiated and planned, using evidence-based management and actual math. It worked. The farms and ranches of the Siskiyou were productive, the valley rich with wildlife and food. The water was controlled to aid the salmon in their travels upstream to spawn, which was easy for the chinook and king salmon, which come in the fall and leave in the spring when the waters are high. Siskiyou County was one of the richest rural counties in the nation until the forests were shut down in 1993. Still the county staggered on, its ranches and farms paying the freight, employing suppliers and service workers, keeping the schools and fire halls and county government open, keeping the county alive.

For the past thirty years, the movement has worked hard to convince the public that all dams are destructive and must be taken out, largely for the sake of the fish but also for the health of the rivers' ecosystems. By the end of 2011, 925 significant dams in the American heartland had been removed. That number does not include the removal of thousands of tiny dams once used to feed and water livestock and small farms up in what are now national forests, wildlife refuges, experimental zones, and formal wildernesses. It hardly needs saying that any privately conserved lands have their dams removed. American Rivers and its satellites want to reconnect rivers to their floodplains and eliminate any channelized ditch, because irrigation is "unnatural." They tout spectacular flood-control benefits from their new water-management system. Because all the dams are not yet removed, the dam system of the United States, one of the most spectacular engineering feats of the twentieth century, is now being managed in order to mimic the natural flows of the rivers in order to return them to pre-Columbian conditions. This is why Chuck Leaf's research is ignored. The Corps of Engineers does not want increased water flow to downstream communities and agriculture. The flood supercharging under the Platte River is deliberate. The movement wants those communities to be flooded out of existence. They are not "natural."[8]

The catastrophic flooding of the Missouri in the summer of 2011 was caused by a refusal on the part of the Corps of Engineers to release water early and in stages from the reservoirs high up in the system in a year when the snowpack was 500 percent of normal in some places. Management that adapted to natural conditions was not natural and therefore not good. Finally the water broke the barriers, the dams were opened, and the result was catastrophic. The Missouri spread out eleven miles at its widest, and entire towns were inundated. Neighborhoods that predated the dams now lie beneath the surface of the water. What American Rivers—and the Sierra Club and all the usual suspects—don't want you to know is that their touted-as-superior "natural flood control" means that no one will be able to live or farm in any floodplain. Their plan, already part effected, will destroy tens of millions of lives, hundreds of billions of dollars' worth of public infrastructure, and a good part of the wealth of the United States by destroying the nation's agricultural bounty.

Siskiyou County is not taking any of this lying down. The removal of the four dams on the Rogue River will devastate the riverbed for sixty years, kill any salmon run for decades, flood and cover thousands of miles with toxic sludge, destroy the inhabitants' way of life, ruin family fortunes, crash property values, and eliminate services for the poor, elderly, and sick, thereby destroying the community, and they know it. They have observed the multipronged attack on their lives and are mimicking it in order to fight back. In the 2010 elections, 80 percent voted against dam removal, and every county commission, fire district, and school board is against dam removal. The high sheriff Jon Lopey became famous for saying that he would not enforce some of the new regulations: "There's no way I am criminalizing the citizens of this county." And they have hired Fred Kelly Grant to help them coordinate. It may be too late, at this juncture; the plan is so advanced that the dam removal merely requires the signature of the secretary of the interior, Ken Salazar (generally called a dirtbag in rural areas), and he has already indicated he is ready to sign.

This will be the largest dam removal in history, and as such is critical for the movement. The Klamath River Basin and its bordering counties are meant to be wild lands, emptied of people and activity. For the Wildlands Network, the Wildlands Project, or the Wild Corridors Project, Siskiyou County is the Ark of the Covenant, and the wished-for monument designation is called the loading ramp of the Ark of the Covenant. These names hold a magnetic resonance for the city people who fund the attack on the people of the valley. What the fund-raisers don't tell their marks is that they are forcing the conformation of the Siskiyou to a UN plan, that Siskiyou County wildlife, grasslands, and forests will die without man's stewardship, and that the land will go to desert within a hundred years.

Dozens of Siskiyou County residents have fanned out, talking to anyone who will listen. Digging into the data, residents, scientists, and engineers show that most of the data is so patently false that it is clear this is a made-up crisis. Busted California is about to pay $1 billion to remove dams for the sake of a fish not common in the area, the coho salmon, which the California Department of Fish and Game kills by the hundreds of thousands every year.

"The government planted the coho here in the early 1900s," says Debbie Bacigalupi. "The Klamath is an inverted basin, the water quality improves as the river flows to the sea. Up in the headwaters, because the source is volcanic, the water is rich in phosphorus and magnesium; it is not a good host for the salmon."

As in every other enviro taking, democratic rights are removed. "Two agreements, one called the Klamath Basin Restoration Agreement and the other called the Klamath Basin Hydro Agreement, mean, once we have signed on—and they are offering half a billion dollars to us to sign on—we won't have any say in what happens in the county. We will lose our voting rights, and our democratic rights," says Bacigalupi. "A committee of 'stakeholders,' consisting mostly of representatives from environmental organizations and federal and state agencies, will decide what happens to our land and water."

Siskiyou County residents, if they can find a way to stay in the county, will be reduced to serf status. Administrative councils are typical environmental reorderings of something we all take for granted—proportional representation and voting—meant to transfer power to environmental organizations, conservation bureaucrats, and their funders. There is always a payout. Three of the four Indian tribes in the region have signed on. "They will be given eighty-seven million dollars and tens of thousands of acres of land,"[9] says Bacigalupi.

Bacigalupi's parents have a two-thousand-acre ranch on the Rogue River. Her father, a construction engineer for the Department of Transportation in Sacramento, bought his ranch midcareer. Jerry Bacigalupi designed a fish bypass tunnel that would solve the issue of the coho needing a route upstream. Dam removal in California will cost $1 billion, and fish ladders would cost $300 billion, but fish tunnels, with no environmental impact because they run underground, will cost $50 million. Fish and Game ignored this low-tech solution.

"Whether the dams go out or stay in, if you read the eighteen hundred pages of the agreement, they are taking our water no matter what. I don't think anything has happened as fast and in so many areas—that we have heard—as in Siskiyou County.

We think they were counting on a poor community that they could plow over. And with that, the government has used our massively progressive-diseased state to attack. We have the largest dam removal in history; they've designated us as a 'biodiversity hot spot,' monumentalization with the 'Climate Refuge' [the Ark of the Covenant], destroying our forests/timber, increasing the ranching/farming regs, the mining/dredging moratorium, road closures, spotted owl, coho salmon, salamander, water, elderberry, juniper tree, dust, hay . . . it's everything. They're killing us.

"What the enviros and the government want to do is called a 'whole basin restoration.' That means all water in Siskiyou County is up for grabs. This means Lake Shastina (and the golf course and community) will be devastated too because the Dwinnell Dam is on the Shasta River and will also have to be removed." Lest the well-heeled upper middle class think that this will never touch them, a gated community filled with million-dollar houses just below the Iron Gate Dam has seen house values drop 90 percent.

John Menke holds a doctorate in range systems ecology and quantitative ecology, the kind that uses actual math. For much of his working life, he worked as a full professor at the University of California–San Diego. He explains coho reality: "The coho were native to the Klamath, but abundance [was] extremely low—sometimes only five hundred returning coho a year—they are typically a coastal spawner. Coho in a healthy population in over two hundred linear miles is extremely unusual; it can happen, but they can't healthily oversummer here, because we have extremely hot summers, and no fog cover, May to November. Coho have never been abundant except within twenty-five miles of the ocean in California, where you have the summer fog belt.

"What people don't know is that in the late 1800s, canneries in the Klamath almost extirpated the fish. They were stopped, and the government built a hatchery that spawned 2.5 million coho, which were raised up here starting around 1895, with the big moves in 1913, before they knew about the coho's limitations with regard to oversummering in our hot, dry Mediterranean climate.

"When they built the Iron Gate Dam in '62, the fuss ended up in the Supreme Court, which decreed that the state would run a hatchery to produce five million chinook, one million steelhead, and seventy-five thousand coho to mitigate the blockage of the river. That very effective hatchery has fostered a pretty good run of coho up here. It waxes and wanes, but the upshot is that the fish has populated many of the rivers and streams up here and provided a basis for this big battle.

"Based on that evidence, which is false evidence, the environmental movement is using this fish as a device and a pry to shut down agriculture. It's a big agenda from the government—they got the fish listed as a separate unit.

"Because of the dams, the hatchery now trucks and hauls the salmon above the dams, but when they come back, if there are more than seventy-five thousand—and there generally are—they electrocute them. That seventy-five thousand returning coho produce four hundred to five hundred thousand fry every year, which is too many, so they electrocute them, too, or heat the water until they're cooked, or poison the water. They kill all but seventy-five thousand fry to release downstream to the ocean. They can triple the number of coho in the hatchery, but they are not allowed to do that. Those poor bastards at the hatchery, their whole life is spent in one meeting after another trying to figure out what to do with a government run amok.

"The lack of fish overall is due to unsupervised, ancestral-rights-based net fishing by Native Americans and overfishing in the oceans. It has absolutely nothing to do with ranching two hundred miles from the ocean. The agencies in here have been lined up by the environmental activists."

In the summer of 2011, Doug Jenner, a fourth-generation rancher, along with his brother, nephews, and sons, were found on their ranch and read their rights because Fish and Game had found five dead coho salmon fry on one of their dry creek beds. The fine was $25,000 per fish. An employee of the National Oceanic and Atmospheric Administration had driven two hundred miles to make an example of the family. With an armed cop in a flak jacket standing nearby, they were cuffed, read their rights, and informed that a decision would be made on whether they would be prosecuted criminally or civilly. This action infuriated Jon Lopey, the head sheriff of the county, so entirely[10] that he put together a coalition of sheriffs from eight counties around the basin to plan legal retaliation against dam removal. The movement spread, and today the high sheriffs of California and Oregon, the final authority in their counties, are united against dam removal.

California Fish and Game may have to call the army into Siskiyou if the environmental movement gets its way. Luckily for them, the U.S. Senate has just passed a bill allowing the federal government to designate its own citizens as terrorists and be detained without trial.[11]

CHAPTER FIFTEEN

What If It's All a Lie?

Surrogate issues steer a hidden agenda everywhere. Luckily they've bungled the hiding, too, and to discover its source, all you have to do is drive down the spine of the continent to a collection of dust-covered, flat-roofed, sand-colored small towns in New Mexico's boot heel.

In Lordsburg, two out of three storefronts are shuttered, and the only thing that has had a coat of paint in the last twenty years is the Hampton Inn, built in the flush of the building boom, redolent with new-hotel off-gassing. The Inn is a palace by comparison to everything else, and I settle down in it, waiting for Judy Keeler to get herself free of the meetings that eat up her days. Keeler is on every committee in the area, and right now she is working on border problems, "right in the heart of the enemy," she says, by which she means the Malpai Borderlands Group, which, in a typical ploy of green groups, works hard to co-opt business leaders like Keeler. It started when the Nature Conservancy bought the keystone ranch in the boot heel, the Gray Ranch; the Malpai Borderlands Group bought off neighboring ranchers by giving them money for conservation easements and promising that they could graze on the Gray Ranch. This arrangement, in typical green fashion, has meant nothing but trouble ever since. And two years ago, the activity of the green activists in the boot heel caused the death of rancher Robert Krentz Jr., who was shot on his own land by an illegal immigrant passing through.

The Border Patrol cannot go onto any conserved lands with mechanized equipment, like a truck, without a permit. This means that 4.3 million acres on the border are virtually unpatrolled; drug smugglers and human traffickers pour through undetected. Nor can the Border Patrol place surveillance cameras in the area or even build communication towers. Last year, after Krentz's death and an almighty fuss, the Border Patrol did manage to build an outpost in the wild lands, but it is twenty miles from the border and located in a depression in the land, which means the necessary line of sight is obstructed. Unless an illegal is captured within five miles of the border, says the Border Patrol, 80 percent will get in and never be found again. Congressional staffers say illegal aliens know exactly where the Border Patrol can and cannot patrol in their vehicles. But the Malpai Borderlands Group wishes to substantially increase that wild acreage to increase habitat for endangered species.

Robert Krentz was a friend of the Keelers. For Judy, the free passage of armed human traffickers and drug smugglers through her counties is yet another life-or-death issue forced upon county residents by environmentalists.

Judy and her husband, Murray, have two ranches, eighty miles apart. The range down here is arid, and you need a lot of land to run two hundred head of cattle. Before the border issue became all-consuming, Judy battled the Wildlands agenda for twenty years, and she is wise to all the machinations of the patrician crew whose ideas have collapsed the economy of her beloved Animas and Hidalgo counties and now threaten the lives of county residents.

The howling wolf of the environmental movement, Dave Foreman, lives in nearby Albuquerque, and the repeatedly renamed core of Agenda 21 and the Global Biodiversity Project, the Wildlands Project, is his pet, the core legacy of the man who is arguably the most destructive radical of the Sixties generation. The Wildlands Project was first renamed the Wildlands Network and replaced in 2011 by the Wild Corridors Project. Its title may go through another iteration before you read this sentence; this tactic typically is used when too much bad press attaches to a green project. New Mexico's boot heel was projected as the core land of the Wildlands Project/Wildlands Network/Wild Corridors Project, and its heart was the Gray Ranch. When first bought by the Nature Conservancy, it covered seventy square miles, and the reserve has now maxed out at five hundred square miles. It was renamed Sky Island, a name resonant with biodiversity bliss and wild and crazy freedom.

TNC sold the Gray Ranch to Drummond Busch Hadley, an heir to the Budweiser fortune. At first the ranch was mothballed, except for upgrading the housing for virtuous rich people so they could commune with the desert. His son, Seth, managed the ranch on a simmer, but is now supposed to be bringing it back. His will be a long, slow crawl. TNC cut the cattle by about 90 percent, and while they do still pay taxes, the county tax assessor had a visit from the Budweiser family lawyers. They threatened that if the assessor ever raised the value of the property, they would exercise their nonprofit status.

Revenues to the county dropped 75 percent from its mainstay ranch. "The next thing that happened was that the county lost the money churn from that ranch, which was considerable. Also the jobs, also the major client for machinery and fodder and so on," says Keeler. "By the time the acquisition of the full five hundred square miles was completed, the economy had deflated by a third. Then the wildlife moved from the Gray Ranch to neighboring ranches for the water, because the Sky Island Alliance quit maintaining all their water windmills. Wilderness doesn't have windmills. When they bought neighboring ranches, they ripped out infrastructure, tore down pioneer houses."

From then on, it was one disaster after another. "TNC and Fish and Wildlife just love fire. They started a burn in the Animas Mountains, from the back side. They decided they didn't want it to get into the grasses because then everything would burn, so they started a back burn at the bottom. These two fires were burning together and we were watching it, and we wondered where the Coues deer would go. They're little and their habitat is very small and they don't like to wander around. We can't prove it, but they killed a bunch of Coues.

"Charlie Painter, a herpetologist, has been studying ridge-nosed rattlesnakes in the Animas Range for twenty years. I asked him what happened to them. He said they survived the fire, went down in the ground. Then the rain came, and because the fire had killed all vegetation, the rains flooded the range, killed the rattlesnakes. And that canyon, you wouldn't even recognize it now. It's a completely different landscape. They did the same thing with the white-sided jackrabbit—a state endangered species. They started a fire on the Fitzpatrick, they let it burn, and the rabbits would come out of the fire so dazed, they'd turn and go back into the fire. The grasses were gone, so the rabbits who survived were easy pickings for the raptors.

"They devastate these areas. And they protect each other. If a rancher had caused all that wildlife devastation, he would have been in court, and fined."

Then, as Foreman's Rewilding America site touts, the crew working in Keeler's counties decided to reintroduce the jaguar. This is called a Pleistocene rewilding, which is the flavor of the month in the restoration business. The Bolson tortoise, which was the most recent attempt to shut things down on the Southwest desert, was deemed a Pleistocene reintroduction. Cattle, said the cabal, hurt tortoise habitat, so ranching must be stopped. Ranchers, wise to just how weak these arguments are, commissioned studies, and it turned out that the Bolson tortoise thrived most where there were cattle.

Keeler laughs. "Ten thousand years ago there were jaguars. There is a population of jaguars a hundred twenty miles south of Douglas, Arizona, in Mexico. We call them the circus population. The Catholic Church used to take circuses around in the 1700s, and after they'd entertain the audience, they'd convert them. Many of the circus animals escaped, and this priest I talked to wrote a book that maintains that this population is actually an escaped population, which makes sense. It doesn't get big and lives at a conflux of three rivers. Jaguars love water and love to fish. For a hundred fifty years there have been jaguars, male, ranging up here, but they never stay—they go back to Mexico. Some hunters brought them in from Belize to hunt, and that's what we think we're seeing. The Sierra Club and the Center for Biological Diversity want to bring them back from extinction and claim they were always abundant down here.

"Alan Rabinowitz, from the New York [Bronx] Zoo, is the leading jaguar scientist in the States. He flew across all three states of so-called jaguar habitat, studied it, said that this was not jaguar habitat. He's stuck to his guns, and confronted the World Wildlife Fund. He told us personally that this would be a living hell for the jaguars down here."

For the past fifteen years, Tim Findley, a *Rolling Stone* and (legendarily) *San Francisco Chronicle* reporter during the Alcatraz/Patty Hearst period, crisscrossed the country telling the stories of the men and women locked into what he called "a twenty-five-year cold war for survival." His stories are detailed, accurate, and deeply researched. Findley, who died in late 2010, lived in Nevada and ranged out from there, and I shared the grinding exhaustion he wrote about as the miles unfurled. But it didn't matter. Whatever the weather, the land is so spectacular, there should be choruses of singing, rejoicing angels banked up to the horizon, extending to infinity. The continent is vast, the stories are everywhere, and if you fly over it, you miss the sheer gorgeousness, the endlessly varying beauty, the emptiness.

The country is empty. There are hundreds of miles where you see just a scattering of houses every once in a while. Half-abandoned towns, broken storefronts, collapsed roofs of farm buildings are the most common evidence of humans. The eyes don't lie, as one fervid environmentalist once told me, trying to argue that my shrinking from clear-cuts was a moral good. The country is empty. You don't really get it until you've driven it, north to south, west to east, diagonally across the mountain states, deep into the high mountain desert, and across the Southwest. You can see that where there have been humans living, the grass is lush, the trees are healthy, the terrain is beautiful. Views of rock and sky are stunning, inspirational and arguably connect one to the eternal, but there is an equal or greater beauty where humans have gardened the earth and produced bounty. What you are looking at in those landscapes is love in action. Those small towns are treasures we are losing, have lost.

The Gordian knot of the countryside mess can be solved with one swift blow of the sword. Property rights must be restored to the individuals who are willing to work their lives away tending that land. The people, the individuals and families, in other words, who want it. Confiscation by government, multinationals, and agents of the megarich—the foundations and land trusts—must be reversed. Otherwise, devastation beckons. History tells us that alienating property rights from those who use the land is a prescription for economic failure, and eventually rebellion. In fact, it is axiomatic: weakly developed property rights are correlated with low rates of economic growth and eventual structural damage to the economy.

We're there right now. Our economy is damaged on a structural level. Every economist recognizes this but few put forward solutions. Joseph Stiglitz, the economist who serves as the darling of the left, argues that we must transition from an industrial economy to a service economy. The root of *service* is the same as the root of *serf*; both derive from the Latin *servus*, slave. Servile. But servile to whom?

What is going to happen to these empty lands, the abandoned houses and barns, the neglected wells and water windmills? What will happen to the once independent and strong people forced into the cities to wait on tables and staff big-box malls like teenagers?

Barbara Peterson describes what's happened in the Klamath since 1,400 family farms were put out of business. "I drive by those farms now, and some are green and flourishing. But the houses are not occupied; they are still boarded up and empty. It's Monsanto or the new farmland trusts, buying up the land and putting in industrial farms, growing food using genetic hybrids, which end by devastating the soil. Bred to be impervious to pests, they get a good harvest the first year, an amazing harvest, less the second, less again the third, and by the fourth, fifth, and sixth, the earth has been stripped of its nutrients and goes dead. It doesn't matter to the carpetbaggers,[1] who abandon the land to bugs and dust. But their seed lives on and travels like an invasive on steroids. Monsanto is allowed to do its own environmental impact statements. They subvert the process that has been forced on us, and careless, leave one disaster after another."

In Canada's boreal forest, ENGOs have made deals with

giant multinational forestry companies, shutting local operators out. What is being created is a series of global monopolies, multinationals that have carved up the earth. In a chess game that would impress the grand masters, NGOs, federal agencies, and multinationals divide up the pie, leaving the consumer to pick up the extra regulatory costs of the resource through higher prices. The U.S. Department of Agriculture has entered a deal with the Brazilian multinational JBS, which has taken over two-thirds of the stockyards and meatpacking industry of the United States and, most believe, will operate all the ranches driven out of business by the movement. JBS USA (formerly Swift) will be the major beef producer left in the country. New Zealand Agritech, a dairy producer, entered Georgia through a collaboration with Georgia's land-grant university. The United States is divided into milk zones, generally consisting of two or three states. Georgia is a one-state milk zone, as is Arizona. When NZ Agritech moved into Georgia in 2000, the state had over eight hundred private dairy farmers. As of June 2011, there were twenty-three dairy farmers left.

Nestlé, the world's biggest food company, floats football-field-size bladders out onto Lake Huron to gather its water and has managed to lower the level of the lake twelve feet. Little wonder that some resource-based multinationals seize any advantage they can and behave badly, before anyone notices, then sign on to NGO regulation when caught. If they become subject to an ENGO attack, before long, they are zombies, unaware they are dead. In the popular press multinationals are described as buying off corrupt politicians in developing nations to advance their industrial interests, whereupon they run roughshod over local social, economic, and environmental needs. This is the approach of a predator. An ENGO approaches its corporate subject as a parasite, weakening it but never killing it outright. First ENGOs isolate the weakest or most exposed corporate entities in an industry, and pit them against each other, trying to score political points in a nonexistent competition. Once they make this bottom-up incursion, they slowly nibble into other corporate entities. All the way through the campaign, they are supported by the popular press. Any advantage that industry gains from being agreeable and funding ENGOs is invariably short-lived. When industry allowed ENGOs to demonize clear-cuts as if they were a crime against humanity, they lost the war in the court of public opinion and not just for forestry. When BP suffered an oil spill, they became outright criminals. But BP, prespill, funded Greenpeace. Although this didn't allow BP to do whatever they wanted, it gave BP cover. Greenpeace left them alone and focused their campaigns against Shell. This is known as "demarketing a brand," and ENGOs are masters of demarketing.

ENGOs add substantial costs to every operation. Are these costs strictly necessary? Probably some are, but most are not. Do these costs mean corners are cut elsewhere, in safety and maintenance? Given the evidence of decline in America's public lands, where there is no money for maintenance left, one imagines this must be the case.

Increasingly investors in big industry are retirement funds—and these organizations are very thin-skinned when it comes to dirty environment or dirty social charges. Demarketing forest products, farmed salmon, fracked gas, or tar sands oil looks like it's consumer driven, but it is really investor driven. Multinationals have to answer to their shareholders. ENGOs bear no liability to anyone. They act with impunity.

What's the next step? Truth Squad president Marti Oakley lives a couple of hours north of Minnesota's Twin Cities. She believes that her county will be the first target—the test case—of Mr. Obama's White House Rural Council, along with rural counties in Iowa and Missouri. The czar-heavy council, formed in June 2011, headed by the U.S. Department of Agriculture, consists of all the usual suspects: Interior, the Forest Service, Fish and Wildlife, the Bureau of Land Management, the Environmental Protection Agency, the Council of Economic Advisers, the Department of Energy, et cetera, but also the FBI, the CIA, Homeland Security, and the Department of Defense. The stated goal is to create strong, sustainable rural communities, a new economic paradigm. Most rural people see it as the death blow, and why, they ask, has it been militarized?

"What they're doing is geomapping all resources and, using the precedent set by *Kelo,*[2] will be able to cite eminent domain if they deem your land could produce more," says Oakley. "The BLM and the EPA will determine whether you are making adequate and beneficial use of water and resources on your land. If they decide you are not, and they *will* decide you are not, they will try to take your land under eminent domain. And they will offer you 'fair market value.' If you should say, 'No, I don't want to do this,' the intent is to bring military to get you off.

"We have the leaked memo out of the Indiana office of the Bureau of Land Management that states the goal is to acquire another two hundred thirteen million acres, and they have a billion dollars to pay for it."

The only tool that works for rural people is an idea introduced in the days of America's founding. "Counties were the most important unit of government then, and the Founders recognized how powerful counties were," says Fred Kelly Grant. If state and federal governments, agencies, NGOs, and corporations are forced to tailor their plans for consistency with local customs, economies, culture, and needs, the intractable puzzle resolves itself. Environmental concerns that affect county citizens are weighed along with economic opportunity and community health and coordinated with state and federal policy, multinational NGOs, corporations, and land-use agencies. Local people are equals at the table. "Coordination can revive rural America," repeats Margaret Byfield.

Grant and his children, along with Margaret and Dan Byfield and the other members of American Stewards of Liberty, in Texas, have started to turn the behemoth around. They have hundreds of successes under their collective belt. They need tens of thousands to make a real difference. Grant sees the seeds of salvation in the executive order establishing the White House Rural Council. Section 4(b) clearly mandates coordination with local groups, as does every other environmental regulatory system. Coordination is the key to making that regulation rational. Coordination is not a silver bullet, but combined with an engaged and informed citizenry, it is the only thing that will work.

But like us conditioned adolescents on Salt Spring, rural people have to stand up, put on their adult selves, and get to work. Until coordination is in use in every rural county, forest, range, and water system, the forests will burn, and wildfires—like the four-million-acre Texas wildfire of 2011—will spread across ranges deprived of their water and choked with dry tinder. Those ranges not burned to cinder will turn to desert. Catastrophic flooding is already a familiar event. With all the dams removed, it will get much, much worse. Finally, it's not about the owl, the salmon, the Ark of the Covenant, Sky Island, or even the water. It's not even about the climate, because desertifying the heartland will only increase warming. It's about that ancient Leviathan trying to find a new way of paying for its bloated, grotesque, and preening self. And money to grow.

Acknowledgments

My first teachers in the vagaries of land use were Salt Spring Islanders, sixth-generation farmer John Wilcox and my neighbor Doug Rajala, both of whom fought the trust all the way to the Supreme Court, thereby gutting their family wealth. The forest lot owners of Galiano, too, taught me about the new repressive regime in the country. One hundred families have owned their lands for twenty years and still are not permitted to build even one house on plots as large as 160 acres, unless they donate 75 percent to a nonprofit entity. This reversion of founding principles annoyed me so entirely I couldn't "let go." Further, I was intrigued. How was such a thing possible in the Americas? Other islanders, Eric Booth, Kimberly Lineger, Larry Pierce, and dozens more, illuminated one or another aspect of this new arrangement of the way we live on the land. One of the many good things about being a journalist is that you hear all the stories, and when they all start to sound the same, well then, you've got something. Therefore, I would like to thank the Islands Trust and its fervent and implacable allies for giving me something so rich and dark I was able to spend fascinated years plumbing its depth.

Rob Scagel from Pacific Phytometric Consultants has built an exhaustive digital library on the work of green activists, their pet foundations, the science they use, the vast amount of money at their disposal, and the state-of-the-art strategic planning and communications they wield to the end of shutting down all goods-producing activity in the country. His work has been invaluable to me. Donna Laframboise and Vivian Krause work at deconstructing the science and financial heft of the movement; their work has been so effective, each has managed to shift policy. I consider them sisters in arms, and by the way, none of us take oil money. The late great Fraser Smith organized my finances so that I could work without a net, seemingly indefinitely. Matthew Watters built my statistical analysis of the money fielded by the movement and the displacements caused.

In the United States, the list is long: Frank and Patricia Penwell, Michael Shaw, Fred Kelly Grant, Staci Grant, Margaret Byfield, Dan Byfield, Norm Macleod of Gaelic Wolf Consulting, Dohn Henion, Chuck Leaf, Gary Howden and Lorna Johnson in Ferry County, Debbie Bacigalupi, Alston Chase, Danielle Linder, Kimmi Clark Lewis, Ron Arnold, and Alan Gottlieb were especially helpful. Christopher Ogden, Bill Phillips, and Susan Roxborough helped me shape the story. My daughter, Tracy Johnson, stepped in a couple of times when I couldn't see the forest for the trees, and hacked a path through the brush with her ruthless intelligence. My family was amused and patient even when I skipped Christmas, winter break skiing, and Easter holidays.

Mostly I have to thank Jamie Craig, who acts in my life like a cool shower on a very hot day. He read countless drafts, and his notes made my writing better and especially more exact, the latter by at least one order of magnitude. He also blithely ignored me when I threw pages down the length of the pink house and fired him. I owe him more than I can describe.

Don Fehr at Trident and Adam Bellow at HarperCollins were brave for taking on a book that flies in the face of all conventional wisdom whether on the right or left; it was sheer pleasure working with them. I am deeply appreciative of their contribution and for the rigorous vetting the HarperCollins team gave this fact- and number-rich book.

Notes

Preface

1. Oliver Seymour Phelps, *The Phelps Family of America and Their English Ancestors* (Pittsfield, MA: Eagle, 1899), St. Catharines Museum and author's collection.
2. Charlotte St. John Oille, private memoir, 1870, National Archives, Canada, and author's collection.
3. *Error cascade* is a term in use by scientists confronting the Anthropogenic Global Warming theory. In its essentials, it means bad science institutionalized by public policy, which then triggers more errors, more bad policy, and so on to infinity.

Chapter 1: We Have Been Fooled

1. Dr. Valerius Geist, wildlife biologist, believed wolves were harmless until he moved to Vancouver Island and found his neighborhood surrounded by a pack. After much careful observation, Geist believed the pack was grooming their prey, testing by coming up to back stoops and decks, killing pets and chickens, before they were all shot. He moved on to a thorough reinvestigation of wolf behavior. Three of Geist's essays are available at http://www.boone-crockett.org/conservation/conservation_geist.asp?area=conservation. The new film *Crying Wolf* describes the devastation wolves have created in the heartland. Here is a CBS-TV report on two women surrounded by wolves in Idaho, as well as a (disputed) photograph of the size wolves are attaining and an *Idaho Statesman* report: http://unexplainedmysteriesoftheworld.com/archives/giant-wolf-epidemic-huge-packs-of-giant-canadian-gray-wolves-are-terrifying-idaho-residents.
2. A copy of this memo can be found on R-CALF's website. R-CALF is one of the two national ranching NGOs. http://www.r-calfusa.com/Trade/property_rights/100900BLMLeakedMemo.pdf. The memo details the private land

acquisitions that will be added to the sixty BLM monument/heritage designations now waiting for the president's signature. Monument designations can be signed into law without the need for congressional approval. In every jurisdiction, these designations are being advanced: http://www.co.san-juan.wa.us/council/displayAgenda.aspx?agendaid=309#home#home.

3. Yasunari Inamura, Tomonori Kimata, Takeshi Kimura, and Takashi Muto, "Recent Surge in Global Commodity Prices—Impact of Financialization of Commodities and Globally Accommodative Monetary Conditions," International Department, *Bank of Japan Review* (March 2011).
4. Hernando de Soto, *The Mystery of Capital: Why Capitalism Triumphs in the West and Fails Everywhere Else* (New York: Basic Books, 2009).
5. Figure extrapolated from a study performed by the Urban Institute and the Rockefeller-founded Environmental Grantmakers Association in 2008. The top fifty ENGOs spent $5 billion in 2003. There are another 26,450 environmental advocacy organizations operating in the United States. I used statistical analysis based on the report's figures to arrive at $9 billion in annual revenues in toto.
6. Macdonald Stainsby and Dru Oja Jay, "Offsetting Resistance: The Effects of Foundation Funding and Corporate Fronts," http://s3.amazonaws.com/offsettingresistance/offsettingresistance.pdf.
7. Donna Laframboise, *The Delinquent Teenager Who Became the World's Top Climate Expert* (Kindle Publishing, 2011). Laframboise eviscerates the methodology of the IPCC, finding their reports riven with error and heavily politicized. "Of the 18531 references in the 2007 Climate Bible we found 5587—a full 30%—to be non peer-reviewed." Of the 5,587 non-peer citations, a grand total of six, or 0.1 percent, were flagged as such.
8. Holly Lippke Fretwell, *Who Is Minding the Federal Estate? Political Management of America's Public Lands*, Property and Environment Research Center (Lanham, MD: Lexington Books, 2009), chap. 4, "Is No Use Good Use?"

Chapter 3: The Big Big Picture: Just How Much Land Has Been Saved, Anyway?

1. Paul Johnson, *The Birth of the Modern: World Society 1815–1830* (New York: HarperCollins, 1991). Pages 202–25 describe the degraded and desperate plight of poor Europeans, particularly Scots, at the end of the eighteenth century.
2. Dan Brockington and Jim Igoe, "Eviction for Conservation," *Conservation and Society* 4, no. 3 (2006): 424–70. In this journal article, Brockington and Igoe

round up much of the peer-reviewed literature on conservation displacements, list sixteen pages of studies on those displacements, and include an attempt to calculate the numbers.

3. Mark Dowie, "Conservation Refugees: When Protecting Nature Means Kicking People Out," *Orion* (November–December 2005), www.orionmagazine.org/index.php/articles/article/161.
4. Mark Dowie, "Enemies of Conservation: Enlightened Conservationists Admit That Wrecking the Lives of Millions of Poor, Powerless People Has Been an Enormous Mistake," *Range* (Summer 2006), www.rangemagazine.com/features-summer-06/su-sr-06-enemies-of-conservation.pdf.
5. Mark Dowie, *Conservation Refugees: The Hundred-Year Conflict Between Global Conservation and Native Peoples* (Cambridge, MA: MIT Press, 2009), xx–xxi, 226.
6. Fretwell, *Who Is Minding the Federal Estate?*, and 2005 National Land Trust Census, www.landtrustalliance.org/land-trusts/land-trust-census and www.landtrustalliance.org/land-trusts/land-trust-census/national-land-trust-census–2010/data-tables; Conservation Almanac, Trust for Public Land, www.conservationalmanac.org/secure; Frederick R. Steiner and Robert D. Yaro, "A New National Landscape Agenda: The Omnibus Public Land Management Act of 2009 Is Just a Beginning," *Landscape Architecture* (June 2009).
7. An extensive discussion of British Columbia's lands under public ownership and conserved lands can be found under "Numbers" on my website, www.elizabethnickson.com, as can a précis of how land in the United Kingdom has been locked down.
8. Rene Bruemmer, "Plan Nord: Jean Charest Says Half of Northern Quebec Will Be Protected," *Montreal Gazette* (February 7, 2012).
9. "Ecuadorian Assembly Approves Constitutional Rights for Nature" (July 10, 2008), Climate and Capitalism, http:/climateandcapitalism.com/?p=479.
10. Tom Bethell, *The Noblest Triumph: Property and Prosperity Throughout the Ages* (New York: St. Martin's, 1998).
11. Adirondack Park Regional Assessment Project (May 2009).

Chapter 4: Crony Conservation

1. More formally, ICLEI has "Special Consultative Status" with the UN Economic and Social Council and coordinates local government representation in the UN processes related to Agenda 21.

2. Jeff Goodson, "The Network, Big Brother, Environmental Espionage and the Nature Conservancy," *Range*, Spring 2003.
3. *Wall Street Journal* (May 24, 1989), quoted in *Public Interest Profiles 1992–1993*, 28; also cited by Ron Arnold, *Trashing the Economy* (Bellevue, WA: Merril Press, 1994).
4. In the Pacific Northwest, graduate students thought they'd found a hitherto undiscovered creature and were well advanced into the listing process—and creating their careers—until country people pointed out that it was the mountain beaver, common as the white-tailed deer. Source: Rob Scagel, Pacific Phytometric.
5. Tim Findley, "'Nature's Landlord': The Story of the World's Most Powerful Environmental Group, the Nature Conservancy," Water Rights & Wrongs, *Range* (Spring 2003), www.rangemagazine.com/pdf/spr03_landlord.pdf.
6. Report of the Staff Investigation of the Nature Conservancy, vol. 1, prepared by the Staff of the Committee on Finance, Charles E. Grassley, Chairman, Max Baucus, Ranking Member, S. Prt 109–7, 131.
7. Joe Stephens and David B. Ottaway, "Nonprofit Sells Scenic Acreage to Allies at a Loss," *Washington Post* (May 6, 2003).
8. Brad Wolverton, "Senators Question Tax Breaks Taken by Donors to Conservation Groups," *Chronicle of Philanthropy* (June 8, 2005).
9. Joe Stephens and David B. Ottaway, "Donors Reap Tax Incentive by Giving to Land Trusts, but Critics Fear Abuse of System," *Washington Post* (December 21, 2003): "Broad data about the reliability of claimed deductions are scarce. But a 1984 IRS study examined 42 deductions for easement donations and determined that all but one appeared inflated, resulting in overvaluations totaling nearly $32 million. According to a GAO report on the study, 'The taxpayers generally overvalued their conservation easement deductions by an average of about 220 percent.'"
10. Henry Lamb, "The Green $$$Scam," *Range* (Summer 2010), http://www.rangemagazine.com/features/summer-10/su10-range-green_scam.pdf.
11. Findley, "Nature's Landlord."
12. Jeremy Pratt, Clearwater Consulting Corporation, "Truckee-Carson River Basin Study, Final Report," Western Water Policy Review Advisory Commission (September 1997), report commissioned by Congress.
13. Regulations imposed on farmers and ranchers seem limitless. More edicts (police power attached) are announced every year: the tagging of cattle, more extensive record-keeping requirements for veterinarians and sale barns, new requirements for identifying horses and poultry, and classification of all farm

vehicles as commercial vehicles requiring commercial driver's licenses—precluding seasonal workers, farm children, and grandpas from working on the farm. Why are family farms dying? Government interference. Only factory farms can afford to comply with all the rules. Little wonder that Big Ag campaigns for more rules. On my island, which is a plush exurban rural redoubt, regulations have recently been put in place that control the amount of dust or stench a property owner can raise; if a neighbor doesn't like your pickle recipe or your manure or you, you're in trouble. Risibly, any noise louder than a normal speaking voice is illegal if it can be heard at the property line.

14. For a ten-minute report from historian, NPR correspondent, and rancher's wife Gail Jenner on just what irrigators experience on a daily basis, watch http://www.youtube.com/watch?v=Cr-e4ySec5Y&feature=share. "People cannot sustain life," says Jenner.
15. Timothy Findley, "Water Rights and Wrongs," *Range* (Spring 2003): 12.
16. Findley, "Water Rights and Wrongs," 11.

Chapter 5: Inch by Inch: Biocentric Command and Control

1. G. Evelyn Hutchison, *Acta Biotheoretica* 18:175, per Google Books; also cited by Alston Chase, *In a Dark Wood* (Boston: Houghton Mifflin, 1995): 99.
2. Simon Levin, *Fragile Dominion: Complexity and the Commons* (Santa Fe, NM: Helix Books, 1999).
3. Allan K. Fitzsimmons, "The Illusion of Ecosystem Management," *PERC Reports*, publication of the Political Economy Research Center, 17, no. 5 (December 1999): 5.
4. J. Michael Scott et al., *Gap Analysis: A Geographic Approach to Protection of Biological Diversity*, Wildlife Monographs no. 123 (Lawrence, KS: Allen Press, 1993).
5. www.takingliberty.us is a Web video that illustrates land takings across the country and explains GAP analysis in more detail.
6. Scott et al., *Gap Analysis.*
7. Fretwell, *Who Is Minding the Federal Estate?*, 70.
8. Wendell Cox, *War on the Dream: How Anti-Sprawl Policy Threatens the Quality of Life* (New York: iUniverse, 2006), 74, figure 5.1. Open space includes Classes 1–3 and privately held land lying fallow.
9. Cox, *War on the Dream*, 71; U.S. Census 2000; U.S. Department of Agriculture, "Land Use, Value, and Management: Major Uses of Land," http://www.ers.usda.gov/briefing/landuse/majorlandusechapter.htm.

10. http://www.nytimes.com/interactive/2011/01/07/us/CENSUS.html?hp, from the 2011 Statistical Abstract of the United States: http://www.census.gov/compendia/statab/.
11. Cox, *War on the Dream*, 71, calculated from Census Canada data.
12. Earth Summit in Focus, UNCED, no. 2 (August 1991); Dixie Lee Ray, *Environmental Overkill: Whatever Happened to Common Sense?* (New York: Harper Perennial, 1994), 4.
13. As did dozens of other countries subsequent to the Rio Earth Summit.
14. Crescencia Maurer, *The U.S. President's Council on Sustainable Development: A Case Study* (1999), http://pdf.wri.org/ncsd_usa.pdf.
15. V. H. Heywood, ed., *The Global Biodiversity Assessment* (Cambridge: United Nations Environment Programme, 1995). Unsustainable activities cited can be found on the following pages: 337, 350, 351, 728, 730, 733, 738, 749, 755, 757, 763, 766, 767, 773, 774, 782, 783, etc.
16. Michael Coffman, "Our Federal Landlord," *Range* (Winter 2012): 44.

Chapter 6: The Vigorous Preening of Our Moral Betters

1. Recently surpassed by John Malone.
2. "Deer Causing Rapid Environmental Change on B.C.'s and Washington's Gulf Islands, University of British Columbia," www.publicaffairs.ubc.ca/2011/01/17/deer-causing-rapid-environmental-change-on-b-c-'s-and-washington's-gulf-islands-ubc-study.
3. What? *What?* The history of the wolf reintroduction in Yellowstone is a cavalcade of error and nonsense. First of all, many of the wolves are not wolves; they have dog blood and were bred at the Ghost Ranch. Those captured elsewhere—from Mexico, Canada, or the MacKenzie Valley—and introduced into Yellowstone first preyed upon everything they could find in the park, then set off into the communities outside the park, and then, like the tracking animals they are, set off to their ancestral homes. The prevalence of dog blood in American wolves is the rationale behind the rumored introduction of Russian wolves. Wolf introduction, according to many rural people, is done on the sly by various groups and conservation bureaucrats. Also see Alston Chase, *Playing God in Yellowstone: The Destruction of America's First National Park* (New York: Harcourt & Brace, 1986), for a full discussion of the wolf question, and in the larger sense, how political management does *not* serve true conservation ends.
4. This is called a Pleistocene reintroduction.

5. U.S. Department of Energy, Loan Programs Office, "Our Projects," http://lpo.energy.gov/?page_id=45; U.S. Department of Energy, Office of Inspector General, Office of Audits and Inspections, Audit Report, Department of Energy's Loan Guarantee Programs for Clean Energy Technologies, DOE/IG-0849 (March 2011).
6. Peter Schweizer, *Throw Them All Out: How Politicians and Their Friends Get Rich off Insider Stock Tips, Land Deals, and Cronyism That Would Send the Rest of Us to Prison* (New York: Houghton Mifflin Harcourt, 2011), 74–104; Rick Dewsbury, "JFK's Nephew Received $1.4bn Taxpayer Bailout for His Struggling Green Energy Firm," *Daily Mail* (November 18, 2011); Eric Lipton and Clifford Krauss, "A Gold Rush of Subsidies in Clean Energy Search," *New York Times* (November 11, 2011).
7. thedailygreen.com.
8. Index of Leading Environmental Indicators (Pacific Research Institute, 1996–2009); Stephen F. Hayward, *2011 Almanac of Environmental Trends* (Pacific Research Institute, April 2011); Greg Easterbrook, *A Moment on the Earth: The Coming Age of Environmental Optimism* (New York: Penguin, 1995).
9. Fretwell, *Who Is Minding the Federal Estate?*, 54. Also Lance R. Clark and R. Neil Sampson, *Forest Ecosystem Health in the Inland West: A Science Policy Reader* (Washington, DC: American Forests, Forest Policy Center, 1995).

Chapter 7: It's Not About the Salmon

1. *Truth Squad* is a blog radio show out of Farm Wars (http://farmwars.info) and *PPJ Gazette* (http://ppjg.me) websites.
2. When asked whether the last-minute revisions to Senate bill 1867 provided a measure of safety, Marti Oakley, of the *PPJ Gazette*, one of the leaders of the rural resistance, said, "When you combine this bill with the Military Commissions Act of 2006, the subsequent Defense Authorization Act of 2007 combined with Patriot acts 1 and 2 . . . they already have granted themselves the authority to tag us all as terrorists, strip us of our citizenship, and try us in military tribunals away from the court system." The last-minute revisions provided her no comfort.
3. U.S. Senate Bill 787: "Waters of the United States means all waters subject to the ebb and flow of the tide, the territorial seas, and all interstate and intrastate waters and their tributaries, including lakes, rivers, streams (including intermittent streams), mudflats, sandflats, wetlands, sloughs, prairie potholes, wet

meadows, playa lakes, natural ponds, and all impoundments of the foregoing, to the fullest extent that these waters, or activities affecting these waters, are subject to the legislative power of Congress under the Constitution."

4. Klamath Basin Restoration Agreement Revised Cost Estimates, June 17, 2011, http://216.119.96.156/Klamath/2011/06/RevisedCostEstimates.pdf.
5. Curtis Hayden, "Is the Rogue River Safe? Presence of Heavy Metals from Dam Removals Harmful to Fish and Humans," *Sneak Preview: The Rogue Valley's News and Review* (June 2/July 1, 2011). While this finding has not received due diligence of the kind performed by Erin Brockovich, or received a thorough scientific analysis, it raises a serious question about the human cost of dam removal. If dam removals were found to be releasing toxic chemicals to downstream communities, they would be stopped, which could trigger an investigation of the removal of 925 other dams. The legal ramifications of this for governments, ENGOs that promote dam removal, and businesses, such as Buffett's PacifiCorp, that acquiesce to dam removal are impressive.

Chapter 8: The Real Green Economy

1. Interview, Harbor Master Richard Young, December 2010, "The Pacific Groundfish Buy-Back Proposal is different. This proposal would reduce the groundfish fleet by about 50%. [The actual goal is a range from 40%–65%. Leipzig, 7, http://www.trawl.org/Archived%20Papers/FINAL.PDF.] This is not enough to eliminate all extra capacity in the fishery, as the SSC [Scientific and Statistical Committee of the Pacific Fisheries Management Council] estimated the fleet would need to be reduced by 60% to 90% to achieve maximum economic efficiency. [SSC, 6.] At this time, a 50% reduction in the Pacific groundfish fleet will result in a viable fishery." Young instituted a $50 million buyback program for fishers taken out of employment. "Buying Back the Groundfish Fleet, July 24, 2001."
2. Federal agencies: National Park Service, Forest Service, Fish and Wildlife Service, National Marine Fisheries Service, National Oceanic and Atmospheric Administration, Federal Emergency Management Administration, Army Corps of Engineers, Environmental Protection Agency, Coast Guard, Bureau of Land Management. California agencies: Coastal Commission, State Lands Commission, Regional Water Quality Control Board, Fish and Game Commission, Department of Transportation, Air Resources Board, Office of Historic Preservation, Office of Planning and Research, Department of Boating and Waterways.

3. Dohn Henion, Del Norte County Counsel, "How to Involve Your Agency's Legal Counsel in the Coordination Process," PowerPoint presentation, Call America Conference, Denver, CO, November 2010. Henion's background information is archived at www.elizabethnickson.com. Also, Anna Sparks, fifth district supervisor of Humboldt County, CA, for twelve years, hearing before the Committee on Resources, House of Representatives, 104th Congress, first session, on the effect that federal ownership and management of public lands and the condemnation and restriction of private property have on local areas, March 2, 1995, Washington, DC.
4. John L. Walker, "Tall Trees, People, and Politics: The Opportunity Costs of the Redwood National Park," *Contemporary Economic Policy* 2, no. 5 (1984). "By 1978, actual visitation to the Redwood National Park was only 39,272 visitor days, or about 4 percent of the estimate made for the fifth year of operation. It should be noted that 1978 was also the year when an inordinate amount of national attention was focused on the park due to its expansion. The three adjacent state parks that were originally to be included in the Redwood National Park actually outdrew the national park visitation rate by 18 percent. The most logical interpretation of the data is that at least 15 percent of the visitors to the state parks did not bother to visit the national park (Redwood National Park Headquarters, 1978)."
5. John Loomis, Richard Walsh, and John McKean, "The Opportunity Costs of Redwood National Park: Comment and Elaboration," *Contemporary Economic Policy* 3, no. 2 (December 1984): 103–7. This paper puts the cost in 1984 at $900 million. Also see Walker, "Tall Trees, People, and Politics." Also, from Sparks's testimony: "The estimated cost of the park, as provided to the Senate in 1968, indicated that the original taking of private lands would cost something under 92 million dollars. By 1981, the total was over 306 million dollars. On Wednesday, January 28, 1987, this total was increased by a U.S. district court panel, which concluded the federal government owed additional compensation to Louisiana Pacific Corporation and Simpson Timber Company. The total, including interest due, was expected to be 770 million dollars. Redwood National Park's total cost now exceeds one billion dollars and is the most expensive of all national parks."
6. PILT is payment in lieu of taxes, meant to make up for lost property tax when land values are eliminated or reduced. Typically counties are paid PILT for three years, but in Del Norte, commissioners have had to go to court for any

payment. Likewise, forestry workers were supposed to be compensated and/or retrained and *were* helped until Congress decided that they had been helped enough.

7. Raven Saks, cited in "How U.S. Land Use Restrictions Exacerbated the International Finance Crisis: An Economic Analysis Brief," rev. ed., *Demographia* (2008), www.demographia.com/db-overhang.pdf. Saks, of the U.S. Federal Reserve, has aggregated the numbers. Antiuse land regulations lead to lower economic growth, and the people who are hurt most are those with lower incomes, who find their housing choices and mobility restricted. And a less prosperous economy cannot provide the subsidy assistance they now need but wouldn't if they still had their land.
8. Ron Arnold, "Overcoming Ideology," in *A Wolf in the Garden: The Land Rights Movement and the New Environmental Debate*, ed. Philip D. Brick and R. McGreggor Cawley (Lanham, MD: Rowman & Littlefield, 1996).
9. Niger Innis, "Stop the War on the Poor. FSC and NGO's: Environmental Mythology" (New York: Congress of Racial Equality, 2011); Tim Bartley, "How Foundations Shape Social Movements: The Construction of an Organizational Field and the Rise of Forest Certification," *Social Problems* 54, no. 3 (2007).
10. Williams Group, *What Non-Profits Say: A Study on the Effectiveness of Communications Training* (Menlo Park, CA: William and Flora Hewlett Foundation, 2011), http://whatnonprofitssay.org/wp-content/uploads/downloads/What_nonprofits_say.pdf.
11. Baird Straughan and Thomas H. Pollak, "The Broader Movement: Nonprofit Environmental and Conservation Organizations, 1989–2005," Urban Institute (2008), http://www.urban.org/url.cfm?ID=411797.
12. Environmental Grantmakers Association Publications: http://ega.org/learn/publications.
13. "In 2008–2009 UNEP received US$233.3 million in earmarked contributions, including counterpart contributions and trust funds directly supporting UNEP's programme of Work. The GC25 approved the indicative level of US$228 million for earmarked support in the current biennium of 2010–2011." http://www.unep.org/rms/en/Financing_of_UNEP/Trustfunds/index.asp. The rest of the UNEP budget comes from other fund sources: http://www.unep.org/rms/en/Financing_of_UNEP/Regular_Budget/index.asp and http://www.unep.org/rms/en/Financing_of_UNEP/Environment_Fund/pdf/2012%20EF%20Pledges%20and%20Contributions.16.02.2012.Web.pdf.

14. "Programme activities on climate change, management of ecosystems, environmental emergencies and disasters, hazardous wastes, assessment of the environment and environmental governance, environmental information and law, cooperation with civil society and youth and special initiatives. Special initiatives: coral reefs, save the great apes, billion tree campaign." www.unep.org/rms/en/Projects/index.asp.
15. Financial Plan for the Period 2009–2012, IUCN World Conservation Congress, October 5–14, 2008, Barcelona, Spain. $100 million plus a year: http://cmsdata.iucn.org/downloads/cgr_2008_17_financial_plan.pdf.
16. Innis, "Stop the War on the Poor"; Bartley, "How Foundations Shape Social Movements"; Peter Foster, "Faust in the Forest," *National Post* (January 12, 2012); letter from Michael J. Marx, Coastal Rainforest Coalition, June 5, 2000. Marx threatens West Fraser Timber with attacks on its market if the company does not conform to the Coastal Forest Conservation Initiative.
17. Quoted in "The EU Budget: Lean and Green," EurActiv, July 11, 2011, www.euractiv.com/en/climate-environment/eu-budget-lean-green-analysis–506457.
18. Implementation of Agenda 21, the Programme for the Further Implementation of Agenda 21 and the Outcomes of the World Summit on Sustainable Development, General Assembly, August 9, 2011, 66th session, http://www.un.org/esa/sustdev/documents/agenda21/english/Agenda21.pdf; http://www.un.org/esa/dsd/resources/res_pdfs/ga–66/SG%20report%20on%20Agenda%2021.pdf. Chapter 33 of Agenda 21 is about financial resources and mechanisms. The implementation cost of Agenda 21 in developing countries was estimated at $600 billion each year, of which $125 billion was to come from developed nations. Of the $125 billion needed, only $67–68 billion was committed at the conference.
19. R. Steven Brown, Executive Director, Environmental Council of the States, State Environmental Expenditures, 2005–2008, March 2008, http://www.ecos.org/section/states/spending and http://www.publicagenda.org/charts/federal-spending-environment.
20. Robert Gordon, "EPA Doles Out Taxpayer Dollars to Environmentalist Activist Groups," Foundry (blog), Heritage Foundation (May 19, 2011).
21. Environmentamerica.org and businessgreen.com (February 17, 2009).
22. "US Sustainable Business Spending 2009–14," Verdantix (October 4, 2010). "Verdantix Critical Moments is a globally-scalable model that sizes, forecasts and describes the future direction of sustainable business spending. This re-

port, focused on the addressable US market, provides sustainability leaders in market-facing and corporate roles with a fact-based analysis of sustainable business budgets, market size and forecast data. Based on real financial data from 1,833 firms with US revenues of more than $1 billion in 2008/09, the analysis finds that spending on 29 sustainability initiatives will grow from $28 billion in 2010 to $60 billion in 2014. Over the 2009 to 2014 period the US sustainable business market will experience a 19% compound annual growth rate." Environmental Protection Expenditures in the Business Sector, Statistics Canada, 2010, Catalogue no. 16F0006X: "Businesses operating in Canada increased their spending on environmental protection in 2008, with total expenditures reaching $9.1 billion, up 5.3% from 2006."

23. Verdantix has only measured the United Kingdom and France. I used equivalents to measure the rest. That said, unless the EU member states resolve their financial indebtedness, this number is likely to be lower.

24. Just north of Seattle, on an island with 6,500 residents, ratepayers commissioned a paper on wetlands that contested the findings of the Washington State Department of Ecology, which was demanding a one-hundred- to three-hundred-foot setback from every wetland, creek, and runoff ditch. A property owner would need a study (costing between $2,500 and $20,000) to cut dangerous trees, dig a pond, build a shed, or plant a rose garden within those setbacks, this on an island with seventy-one residents per square mile, many of them present only in the summer. It was a massive property taking that was unnecessary, because newer science believes that imposing setbacks only from productive salmon creeks and wetlands that recharge aquifers is adequate. That commissioned paper ran to 43,500 words and must have eaten up a quarter of the annual budget of the San Juan Ratepayers Association. For that paper to be conclusive enough to turn back regulation, it would have had to have been backed by several dozen specific studies in all the various aspects of hands-on conservation. Essentially the paper pointed out that every study the regulation was based on was not real science, had no science backing it up, but was based on assumptions that were themselves based on assumptions. Nevertheless, the San Juan Ratepayers Association would have had to do forensic data mining to disprove the Department of Energy's "science" and then commission actual science. Cost: $150,000. Time: six months minimum. But for the green machine, countering that one paper meant the flick of a hand, or a few dozen clicks of the mouse as 743,000 words are cut and pasted from other

documents sourced from Timbuktu to Santa Cruz and sent as opposition-flattening fodder to every individual, group, or agency interested in imposing restrictions.

25. In one county of 7,500 people in northeastern Washington, the outfits that moved in and shut down their economy were as follows: Alliance for the Wild Rockies, Colville Indian Environmental Protection Alliance, Columbia Region Biodiversity Campaign, several groups called Forest Watch, Inland Empire Public Lands Council, Kettle Range Conservation Group, Northwest Ecosystem Alliance, Okanogan Highlands Alliance, Pacific Coast Biodiversity Project, Rest the West, Sierra Club Cascade Chapter, and Washington Environmental Council, as listed in "Battered Communities: How Wealthy Private Foundations, Grant-Driven Environmental Groups, and Activist Federal Employees Combine to Systematically Cripple Rural Economies" (Bellevue, WA: Center for the Defense of Free Enterprise, n.d.), www.eskimo.com/~rarnold/Battered%20Communities.pdf, 20. Residents, however, trace all their trouble to two men, who together and separately coordinated and, in some cases, were the only members of the above outfits. The movement likes to create paper groups all staffed by the same people.
26. Paul Allen, cofounder of Microsoft, was the lead donor.
27. Associated Press, "Firefighters Make Progress on Loomis Forest Blazes," *Seattle Post-Intelligencer* (August 6, 2006), www.seattlepi.com/default/article/Firefighters-make-progress-on-Loomis-forest-blazes-1211417.php.
28. Fretwell bases her analysis on the activities of the Land and Water Conservation Fund (LWCF), a leading source of funds for land acquisition. From 1965 to 2002, the LWCF spent $12.5 billion on acquisition and $224 billion on managing the land. Fretwell, *Who Is Minding the Federal Estate?*, 89. Private owners bear the cost of letting their lands degrade, and are therefore motivated. Government, ENGOs, and land trusts do not. The degradation cost of that conserved land is borne by future generations.
29. Although Conservation Northwest raised money to buy out the school trust, which depended on the timber revenue from the Loomis Forest, there still won't be a school if there are no parents earning money.

Chapter 9: It's Not About the Spotted Owl

1. The reopened gold mine cost $100 million in environmental assessments over twenty years, according to Washington state representative Joel Kretz.

2. Forest Policy analyst Jim Peterson is cofounder of Evergreen Foundation and publisher of *Evergreen* magazine, http://evergreenmagazine.com/pages/About_Us.html.
3. A discussion of rural displacements can be found on www.elizabethnickson.com.
4. Fretwell, *Who Is Minding the Federal Estate?*, 80, citing Government Accountability Office, "Wildland Fire Management," GAO-03-805 (Washington, DC, 2003), 14. "According to the National Interagency Fire Center 2000, federal lands made up 62 percent of the burned lands in 2000, even though 80 percent of forests in the nation are privately owned." Fretwell, *Who Is Minding the Federal Estate?*, 81. Also, Douglas MacCleery, assistant director of planning for the Forest Service, says, "The twin problems of fuel build-ups and declining forest health, and their effect on ecosystem diversity and sustainability, are likely to be the single most significant environmental challenges facing federal forest managers over the next two decades." Government Accountability Office, "Reducing Wildfire Threats: Funds Should Be Targeted to the Highest Risk Areas," GAO/T-RCED-00-296 (Washington, DC, 2000).
5. Fretwell, *Who Is Minding the Federal Estate?*, 55, citing *Evergreen*, no. 53, 1994–1995.
6. Fretwell, *Who Is Minding the Federal Estate?*
7. Bond is using the average of four service and related jobs as a multiplier—mainly jobs in machine shops, with suppliers, and in all the many businesses that service the forest sector, as well as the common service jobs that keep a community going.
8. Former county commissioner Joe Bond, as quoted, makes the point that after Boggins Mill closed, family wage jobs were lost, and people who needed a family wage were forced to leave the county. Retirees have moved in—as detailed by the USDA report already cited. Also, rural researchers have found that county populations have shot up because recent immigrants and refugees have been moved into those counties. In an inelegant but descriptive phrase, one rural researcher called the phenomenon "Somali dumps," which is not to say that the rural resistance is racist; quite the contrary. They object to the culture of their counties being overwhelmed and destroyed. In St. Cloud County, Minnesota, the proportion of refugee and recent immigrant to generational farming family is 7 to 1.
9. In British Columbia, 1.3 billion cubic meters or 44 million acres of forest was killed by the pine beetle epidemic. Scientists found a pesticide, MSMA, mono-

sodium methyl arsenate, an organic compound less toxic than other forms of arsenic, which could have worked in tandem with other compounds to kill the beetle and, if used soon enough, could have sharply curtailed the epidemic with limited negative results. However, even thorough testing of the compound was stopped. In Environment Canada's "A Case Against Arsenic-Based Pesticides," www.ec.gc.ca/envirozine/default.asp?lang=En&n=B9657723-1, the authors note that the treatment kills the infected trees and go on to say, "While it was initially believed that MSMA was over 90 per cent effective at killing the beetles, the Environment Canada–led team found the success rate to be only about 60 per cent." The authors go on to cite data indicating a strong possibility of debilitating arsenic poisoning in various animal species. Please note the words "strong possibility." Environment Canada has the reputation of Bernie Madoff to anyone who has driven through the utter devastation of the beetle-kill forest. There is no wildlife in the dead forest, nor is there likely to be for many years. Wildlife has migrated to the healthy forest, if it could do so. If the epidemic had been stopped, even by 60 percent (which itself is a largely untested assumption), perhaps the beetle would not be advancing on the forests of Montana, Idaho, Colorado, or northeast Washington state. An additional two million acres of beetle-kill forest has burned, because the dead trees are dry as tinder. Source: Rob Scagel, forest physiologist, Pacific Phytometric Consultants. The legal antagonist who effectively stopped thorough testing of the compound was a group of physicians out of Smithers, BC, with a relationship to the Sierra Club.

10. "Simulated Reserve and Corridor System to Protect Biodiversity," Environmental Perspectives (Bangor, ME), http://shop.epi-us.com/Maps_c5.htmhttp:/shop.epi-us.com/Maps_c5.htm, widely available on the Web in low-resolution digital form, for example, at www.afn.org/~govern/wildlands.html.
11. It is interesting to note here that people in nearly every rural region I visited think that the movement has selected them as a test case to advance its most destructive initiatives.
12. The International Union for the Conservation of Nature devised a radical remaking of the way rural lands were to be used, formalized in the Biodiversity Treaty in 1988, which the U.S. Senate refused to ratify in a 95–0 vote in 1994. The Wildlands map was instrumental in influencing the vote; property rights would have been vitiated, which violated the Constitution. Rural people swarmed their congresspeople. Nonetheless, in the intervening twenty-five years, the map has been implemented.

13. ICBEMP is the Interior Columbia Basin Ecosystem Management Plan. It was a one-size-fits-all scheme that would have sharply curtailed activity on 144 million acres of public and private lands in the U.S. Northwest. It was abandoned after citizens pointed out that it violated the Constitution, that is, the right to enjoy property. Initial planning costs: $50 million. By 1998, it was dead. But then, so were many of the forest communities of the Northwest.
14. Capp's Senate testimony: www.undueinfluence.com/diana.htm.

An Abundance of Frogs, Snakes, and Water

1. Conservation regulation is so complex and new that each aspect required the services of a separate lawyer who was familiar with that particular aspect of rule making. That way I didn't have to pay for their various learning curves. So I had a highways-covenant lawyer, a density-transfer lawyer, a lawyer for the Natural Area Protection Tax Exemption Program covenant, a water-covenant lawyer, and a lawyer who arranged the loan for my subdivision, which, because the law was untested, required extra documentation and more signing in blood.

Chapter 10: Where Are All the Corpses?

1. Chris D. Thomas et al., "Extinction Risk from Climate Change," *Nature* 427, no. 8 (January 2004): 145–8.
2. Arne Naess, *The Selected Works of Arne Naess* (Heidelberg: Springer Science+Business Media, 2005), vol. 1, 326: "These normative postulates are value statements that make up the basis of an ethic of appropriate attitudes toward other forms of life, an ecosophy." Naess, 1979: "They are shared, I believe, by most conservationists and many biologists, although ideological purity is not my reason for proposing them." David R. Keller, ed., *Environmental Ethics: The Big Questions* (West Sussex, UK: Wiley-Blackwell, 2010). This quote is cited by Michael E. Soulé in the essay "What Is Conservation Biology?" Michael E. Soulé, "What Is Conservation Biology? A New Synthetic Discipline Addresses the Dynamics and Problems of Perturbed Species": "Examples of these supposed collective norms are: diversity of organisms is good, ecological complexity is good, evolution is good, and biotic diversity has intrinsic value." Cited by Monique Borgerhoff and Peter Coppolillo in *Conservation: Linking Ecology, Economics, and Culture* (Princeton, NJ: Princeton University Press, 2005), 67.
3. Chase, *In a Dark Wood,* 108.
4. Laframboise, *The Delinquent Teenager Who Has Become the World's Climate Expert.*

5. Willis Eschenbach, "Where Are All the Corpses?" Watts Up with That (April 1, 2010), http://wattsupwiththat.com/2010/01/04/where-are-the-corpses.
6. Jennifer Marohasy, "Australia's Environment Undergoing Renewal, Not Collapse," *Energy and Environment* 16, no. 3–4 (2005), www.ipa.org.au/library/EE%2016–3+4_Marohasy.pdf.
7. None of this "science" receives anonymous, independent peer review. Typically it receives what is called "over-the-cubicle" review, or review from colleagues who could profit from positive findings.
8. Eschenbach, "Where Are All the Corpses?"
9. M. David Stirling, *Green Gone Wild* (Bellevue, WA: Merril Press, 2008), 137–59; also see Randy T. Simmons and Kimberly Frost, "Accounting for Species: The True Costs of the Endangered Species Act" (Bozeman, MT: Property and Research Center, 2004). Also see the work of Jason F. Shogren, Stroock Distinguished Professor of Natural Resource Conservation and Management and Professor of Economics at the University of Wyoming, especially, Gardner Brown Jr. and Jason Shogren, "Economics of the Endangered Species Act," PERC Series, PS-3, and *Journal of Economic Perspectives* 12, no. 3 (Summer 1998): 3–20.
10. Karen Budd-Falen memo, September 30, 2009, http://www.buddfalen.com.
11. Tom Knudson, "Litigation Central: A Flood of Costly Lawsuits Raise Questions about Motive," *Sacramento Bee* (April 24, 2001).
12. J. P. Freire, "Report: Department of Interior Employees Collaborate with Lobbyists," *Washington Examiner* (October 5, 2009), http://washingtonexaminer.com/politics/beltway-confidential/2009/10/report-department-interior-employees-collaborate-lobbyists. From the report: "Our investigation determined that numerous activities and communications took place between NLCS officials and nongovernmental organizations (NGO) including discussions about the NLCS budget and BLM employees' editing brochures and producing fact sheets for a specific NGO. Our investigative efforts revealed that communications between NLCS and certain NGO's in these circumstances gave the appearance of federal employees being less than objective and created the potential for conflicts of interest or violations of law. We also uncovered a general disregard for establishing and maintaining boundaries among the various entities."
13. Knudson, "Litigation Central."
14. Karen Budd-Falen, "Western Legacy Alliance Research Regarding Federal Government Funding of Environmental Litigation against the Federal Government" (Cheyenne, WY: Budd-Falen Law Offices, October 22, 2009), http://

www.buddfalen.com/eaja.html. In this quote, Budd-Falen cites Tony Davis, "Firebrand Ways," *High Country News*, www.hcn.org/issues/41.22/firebrand-ways/print_view. *Utne Reader* credits this as the original source at www.utne.com/blogs/blog.aspx?blogid=36&tag=endangered%20species.

15. Coyotes are protected in twelve states.

Chapter 11: The Country Is Being Liquidated

1. Conservation easements remove property rights from land in order to conserve it, thereby depreciating the land's value. Easements typically have been access easements, like roads through one property to another property or access to power or water lines. Conservation easements have been shown in some academic studies to give their owners seniority on the land.
2. Between 2000 and 2008, the Nature Conservancy received $317 million from the U.S. government for land acquisition. IRS Form 990.
3. Ed Depaoli, former area manager, Bureau of Land Management, "Response to RangeNet," *Range* (Spring 2001).

The Ehring Bunker

1. The signatures and notes can be read at http://www.ipetitions.com/petition/review_islands_trust_act/signatures.
2. Bruno Ganz as Hitler in a clip from *Downfall* (*Der Untergang*), 2004, www.youtube.com/watch?v=t7PmzdINGZk.
3. "The Ehring Bunker," www.youtube.com/watch?v=MDcrSJ2YObw.

Chapter 12: High Noon in Owyhee: Rebellion Begins

1. It passed. Not only did it pass, but it gave Monsanto permission to perform its own environmental assessments, despite the fact that a good deal of evidence shows that some of Monsanto's genetic hybrids act as aggressive invasive species that prey upon neighboring farms and destroy traditional seed stocks.
2. Wallace Stegner, *Beyond the Hundredth Meridian: John Wesley Powell and the Second Opening of the West* (New York: Penguin, 1954); Wayne Hage, *Storm over Rangelands: Private Rights in Federal Lands* (Bellevue, WA: Merril Press, 1991).
3. Mountain Home Air Force Base is in Owyhee County.
4. Every other deal—like the much-touted Quincy Library and Steens Mountain arrangements—has foundered in a mess of litigation, with no resolution in sight.
5. Deborah Epstein Popper and Frank J. Popper, "The Great Plains: From Dust

to Dust," *Planning* (December 1987), www.planning.org/25anniversary/planning/1987dec.htm.

6. Wayne Burkhardt, "How the Desert Blossoms without Livestock Grazing," *Range* (Spring 1993).
7. John L. Schmidt and Douglas L. Gilbert, *Big Game of North America: Ecology and Management* (Harrisburg, PA: Stackpole Books, 1978).
8. Factor this in. Fish and Game placed a screen where the lake decants into the ditch system to prevent bull trout from getting out of the lake and into the irrigation system, so that when the system goes dry, as those with no point-source water do when water isn't needed for cattle, the trout wouldn't die. But the gaps in the screen—*designed by Fish and Game*—were too big, and bull trout fry were escaping into the ditch system and dying. So Fish and Game tried to force the irrigation district into getting an "incidental take" permit, asking permission to "take," that is, kill, an endangered species. The irrigation district refused, because the permit would mean they were confessing to taking an endangered species, and down the road they risked having their system shut down and themselves fined, and fined substantially. The local Fish and Game official informed the irrigation district that, given that they wouldn't apply for a take permit, the water in their irrigation district channels would have to be kept at a certain height, with a certain minimum stream flow, and that flow rate and depth took precedence over irrigating ranch land. Miles of irrigation ditches must now be monitored to save a species that (a) did not exist in the irrigation system and (b) shouldn't be sheltering in a man-made ditch that waters thousands of acres of ranch land and heads of cattle, which support the economy of the region.
9. Owyhee County was Marvel's first attempt to—as he affirms repeatedly—destroy the ranching economy of the West. He failed with the Owyhee but has received tens of millions of dollars from corporations and foundations that support his goals. At no time has any independent audit been performed on the millions of acres of ranch land mothballed by Marvel. Little doubt those acres are turning to desert. Marvel would have some healthy demonstration ranges to show to his funders, but an independent audit of the millions he has shuttered would be instructive.
10. The Yellowstone to Yukon map can be viewed at http://maps.y2y.net/googlemap.htm.
11. Despite the many researchers who have followed in Arnold's wake, he still has the best grasp on the weight and measure of the movement.

Buffering Paradise

1. Photos and description of rammed-earth building, www.elizabethnickson.com, Pink House Project.
2. Legal nonconforming sounds comforting, unless you have seen, as I have, in-house presentations by conservation bureaucrats showing the steps by which legal nonconforming structures can be turned into illegal nonconforming ones and finally eliminated.
3. Scientists largely attribute the bisexing of amphibians and fish to birth control and other hormones flushed from women's bodies into the water system. See A. L. Filby et al., "Health Impacts of Estrogens in the Environment, Considering Complex Mixture Effects," *Environmental Health Perspectives* 115, no. 12 (December 2007); K. A. Kidd et al., "Collapse of a Fish Population after Exposure to a Synthetic Estrogen," *Proceedings of the National Academy of Sciences* 104, no. 21 (2007): 8897–901; K. E. Liney et al., "Assessing the Sensitivity of Different Life Stages for Sexual Disruption in Roach (*Rutilus rutilus*) Exposed to Effluents of Wastewater Treatment Works," *Environmental Health Perspectives* 113, no. 10 (2005):1299–307; D. Martinove et al., "Environmental Estrogens Suppress Hormones, Behavior and Reproductive Fitness in Male Fathead Minnows," *Environmental Toxicology and Chemistry* 26, no. 2 (2007): 271–78, as cited in Iain Murray, *The Real Inconvenient Truth* (Washington, DC: Regnery, 2008).
4. Problems are place specific, demanding specific place-based amelioration.
5. Donald F. Flora's MS and PhD from Yale are in forestry and geology. He has forty years of research experience in the natural sciences. He was in charge of several forestry research laboratories in the Northwest, in Oregon and Alaska. He was technical editor of the *Journal of Forestry* and the head of National Fire Danger Rating System research. He is also the former head of the National Timber Harvest Issues Program and a former affiliate professor at the University of Washington. He's a former director of Keep Washington Green Association, with an eighty-year family history of experience in Puget Sound shoreline ownership and stewardship. He is currently reviewing 3,500 research papers on buffers, riparian zones, beach functions, and fisheries.
6. "They said, 'We will design buffers which are 1 SPTH—site potential tree height.' An old-growth tree will be two hundred feet tall at four hundred years old. If that tree falls down, it falls downhill into the stream. Foresters used to clean out streams, but biologists asserted woody debris fed creeks. Then they

decided it would be one hundred fifty feet. Or sometimes, 1.5 tree heights or two hundred twenty-five feet on each side." Interview with Don Flora, April 2011.

7. Harvests are down to 20 percent of 1980 harvesting level in all U.S. forests and between 1 and 10 percent of the harvest in the forests designated as spotted owl habitat since the listing of the owl.

Chapter 13: Leviathan Cloaked in Darkness

1. UN Conference on Human Settlements, May 31–June 11, 1976, Vancouver Action Plan, part D, Land, http://habitat.igc.org/vancouver/vp-d.htm.
2. MacBride presided over the UN's International Commission for the Study of Communication Problems, which in 1980 issued its statement of purpose: Seán MacBride et al., *Many Voices, One World*, rev. ed. (Paris: UNESCO, 1988; Lanham, MD: Rowman & Littlefield, 2003), widely available online, e.g., at http://ics.leeds.ac.uk/papers/vp01.cfm?outfit=ks&folder=4&paper=7.
3. Peter Brimelow, "Stalking the Boy Wonder," *Maclean's* (October 4, 1976).
4. A number of citizen-scholars have taken the vast rafts of verbiage produced by the United Nations and plumbed those documents without succumbing to blind paranoia or running shrieking into the outback. They've then summarized those documents in workmanlike prose. Principally I drew upon the work of the late Henry Lamb. Michael Shaw and Michael Coffman have built upon his work, both speaking to ongoing issues raised by UN incursions or American sovereignty.
5. *Global Biodiversity Assessment* (Cambridge, UK: Cambridge University Press, 1996), section 11.2.3.2, 773.
6. The Tides Foundation is notorious for its secret cadre of donors who give lavishly but do not want their donations made public. Often called a money-laundering outfit for rich socialists, donors include the Heinz, Turner, Pew, Rockefeller, Hewlett, and AT&T foundations, as well as George Soros's Open Society Institute, the Rainforest Alliance, Barbra Streisand, and others. See http://activistcash.com/organization_overview.cfm/o/225-tides-foundation-tides-center and www.discoverthenetworks.org/funderProfile.asp?fndid=5184.
7. Kenneth Anderson, "Agency Failure, Legal Liability and Jurisdiction, and International Organizations," Volokh Conspiracy, September 16, 2010, http://volokh.com/2010/09/16/agency-failure-legal-liability-and-jurisdiction-and-international-organizations.
8. Macdonald Stainsby and Dru Oja Jay, "Offsetting Resistance: The Effects of Foundation Funding and Corporate Fronts from the Great Bear Rainforest to

the Athabasca River." A chart of how money in a campaign is marshaled is included in this report, which analyzes how private foundations work with corporations to control a resource industry and shut out smaller operators. Stainsby and Jay work from the perspective of the Left and snap in an important piece of the puzzle, http://s3.amazonaws.com/offsettingresistance/offsettingresistance.pdf, 12–13.

9. David Gritten and Blas Mola-Yudego, "Blanket Strategy: A Response of Environmental Groups to the Globalising Forest Industry," *International Journal of the Commons* 4, no. 2 (August 2010): 729–57. Other scholarly analyses of the cooperation among ENGOs, multinationals, and senior governments that alienate resources from a country's people are cited in Gritten and Mola-Yudego's paper.
10. Tim Findley and Diane Alden reported on the story as it unfolded. In this case, Diane Alden, "Wayne Hage's War: How the Monitor Valley Adjudication Came to Be," *Nevada Journal* 6, no. 4 (April 1998), http://nj.npri.org/nj98/04/hage.htm. Also Tim Findley, "David and Goliath," *Range* (Summer 2001) http://www.rangemagazine.com/archives/stories/summer01/david_goliath.htm.

Victory

1. As of this writing, the case is under appeal. The trustees were acquitted of conflict of interest. However, the Supreme Court judge closed his remarks as follows: "Finally, even though I have found that the petitioners have failed to establish peciuniary or non-pecuniary conflict of interest in this instance, I note that many of the issues raised by the petitioners may fall in fact under another aspect of natural justice, namely statutory standards for procedural fairness . . . the petition asserts that the respondents contravened the open meeting provisions of the community charter, yet—as counsel for the petitioners properly acknowledged—the petitioners ask for no relief under this heading. The petitioners might very well have argued that the manner of the notion and vote could invoke a finding of procedural irregularity which could result in the vote being set aside."

Chapter 14: It *Is* All About the Water

1. The Wildlands Project/Network/Wild Corridors Project aims to create the North American section of the UN's network of protected areas. The names change with regularity. Ron Arnold calls those changes disambiguation, which is a nice word, perfectly explaining the shifting images the movement uses whenever criticism hits home. http://www.undueinfluence.com/wildlands_disambiguation.htm.

2. A typical family uses one acre-foot a year; in arid regions, families get by on one-fourth acre-foot.
3. The theory was first advanced in 1912, in its essentials proposing that a forest should be managed to be open enough to allow snow to reach the ground and dense enough to protect the pack from the sun and wind. That would produce optimal snow conservation. Leaf has refined that theory, tested it extensively in the Fraser Experimental Forest, and increased water yield 35 percent. With those results, Leaf then created exact metrics for the cuts and tested them.
4. Steven Neugenbauer, LG, LEG, LHG, PG, RG, REA. His credentials translate as follows: Licensed Geologist (Washington), Licensed Engineering Geologist (Washington), LHG Licensed Hydrogeologist (Washington), Professional Geologist (Wyoming), Registered Geologist (Kentucky), Registered Environmental Assessor.
5. According to the Washington State Department of Ecology, 75 percent of the toxic chemicals that flow from Seattle into Washington's Puget Sound are carried by storm water that runs off paved roads and driveways, rooftops, yards, and other developed land.
6. The Klamath Tribes Sociocultural/Socioeconomics Effects Analysis Technical Report, U.S. Department of Interior, Bureau of Reclamation, Technical Service Center, Denver, CO, http://klamathrestoration.gov/sites/klamathrestoration.gov/files/EIS-EIR-Draft/Econ-Reports/Tribes/Klamath%20Tribes_9-16_FULL.pdf.
7. Briefly, and it is possible to go on at book length (as it has been) with regard to the twisting of truth in the Klamath Basin. A few of Peterson's "lies" follow. That September, low warm water led to the deaths of some 70,000 adult chinook returning to the Klamath to spawn, according to the California Department of Fish and Game. The Klamath Basin irrigators were blamed by NGOs and agencies for that fish kill, claiming that the "low flows" were caused by their use of the water. But in 2002 there were studies done on the meth labs on the Klamath that drained into the Klamath River. These reports were not cited as a factor in the fish die-off. Also mentioned at a science workshop in Klamath Falls was the possibility of the months of black skies from the summer of wildfires creating the die-off. Fires put ashes into the river and robbed it of oxygen. These were not considered either. Finally, the National Research Council's final peer-reviewed report said: "In 2002, flows in late summer of 2002 were not atypically low or historic lows. . . . [N]o obvious explanation of the fish kill based on unique flow or temperature

conditions is possible," and "It is unclear what the effect of specific amounts of additional flow drawn from controllable upstream sources (Trinity and Iron Gate Reservoir) would have been. Flows from the Trinity River could be most effective in lowering temperature." Sea lion predation, ocean conditions, entire watershed conditions are factors in fish die-offs as well. None of these was considered.

8. Joe Herring, "Green Ideology Trumps Flood Control," *American Thinker* (August 17, 2011); Joe Herring, "The Purposeful Flooding of America's Heartland," *American Thinker* (June 22, 2011).
9. Klamath Basin Restoration Agreement, Revised Cost Estimates, June 17, 2011.
10. Federal law requires that any federal agent must first announce his presence in the county to the sheriff of that county.
11. According to the *Huffington Post*, despite the last-minute revisions, the provision of S. 1867, or the National Defense Authorization Act, "would deny suspected terrorists, even U.S. citizens seized within the nation's borders, the right to trial and subject them to indefinite detention. The lawmakers made no changes to that language." http://www.huffingtonpost.com/2011/12/13/national-defense-authorization-act-ndaa-obama-detainee-policy_n_1145407.html.

Chapter 15: What If It's All a Lie?

1. Questions abound regarding Monsanto's Roundup Ready crops. All by itself glyphosate could be said to have triggered the organic revolution; people are rightly suspicious. We don't know with any confidence the long-term results of Roundup and Roundup Ready seeds on the soil. Peterson is reporting from the front lines of the "experiment," as it were. But there is enough scholarly evidence to support her claim of glyphosate stripping the soil of nutrients and life, so that it remains an open question. A sampling: Dr. Eva Sirinathsinghji, "USDA Scientist Reveals All: Glyphosate Hazards to Crops, Soils, Animals and Consumers," Farmwars.info; Don M. Huber, Emeritus Professor, Purdue University, APS Coordinator, USDA National Plant Disease Recovery System, "Emergency! Pathogen New to Science Found in Round Up Ready GM Crops," letter to Tom Vilsak, secretary of agriculture, February 21, 2011, http://www.i-sis.org.uk/newPathogenInRoundupReadyGMCrops.php; see also Feng-chih Chang, Matt F. Simcik, and Paul D. Capel, "Occurrence and Fate of the Herbicide Glyphosate and Its Degradate Aminomethylphosphonic Acid in the Atmosphere," *Environmental Toxicology and Chemistry* 30, no. 3 (March 2011): 548–55.
2. *Kelo v. City of New London*, 545 U.S. 469 (2005).

SELECTED BIBLIOGRAPHY

In addition to the sources given in the notes, I found the following to be helpful. A detailed bibliography with 1,200 additional citations, many hyperlinked, is archived on my website, www.elizabethnickson.com.

Chapter One

Butler, Lem, Bruce Dale, Kimberlee Beckmen, and Sean Farley. "Findings Related to the March 2010 Fatal Wolf Attack Near Chignik Lake, Alaska." Alaska Department of Fish and Game, Department of Wildlife Conservation, December 2011.

Cox, Wendell. "The Housing Crash and Smart Growth." National Center for Policy Analysis, Policy Report No. 335, June 2011.

"Extreme Poverty Is Now at Record Levels, 19 Statistics About the Poor That Will Absolutely Astound You." The Economic Collapse, Zero Hedge, November 5, 2011.

Krause, Vivian. "Oil Sands Money Trail." *National Post*, January 18, 2012. Shows how American private foundation money supposedly used to advocate for environmental health is actually used to target competitors and to enhance U.S. geopolitical power (which obviates their tax status).

Post Reporters. "Global Food Crisis, the New World of Soaring Food Prices," series. *Washington Post*, April 2008.

Stanbury, William T. *Environmental Groups and the International Conflict Over the Forests of British Columbia, 1990 to 2000.* SFU-UBC Centre for the Study of Government and Business, Vancouver, Canada, 2000.

"What's Left of the Middle Class Is More Diverse, Harder Working, and Still Shrinking." *Advertising Age*, October 16, 2011.

Chapter Two

Bulholzer, Bill. "The Islands Trust: 30 Years of Protecting B.C.'s Gulf Islands." *Digest of Municipal and Planning Law*, March 2005.

"Designing the Program." Western Climate Initiative. http://westernclimateinitiative.org/designing-the-program.

Dunster, Kathy. "75 Steps to Preserving and Protecting Paradise: A Slow Islands Action Plan for the Islands in the Salish Sea." Dunster and Associates, Environmental Consultants, Ltd., September 18, 2005.

Kemmis, Daniel. *Community and the Politics of Place*. Norman: University of Oklahoma Press, 1990.

Chapter Three

Coffman, Michael S. "Special Report: Our Federal Landlord." *Range*, Winter 2012.

Coffman, Michael S. "Taking Liberty: How Private Property in America Is Being Abolished." *Range*, Fall 2005.

Dickenson, John. *Letters from a Farmer in Pennsylvania to the Inhabitants of the British Colonies*. London: John J. Almon, 1768.

Driessen, Paul. *Eco-Imperialism: Green Power, Black Death*. Bellevue, WA: Free Enterprise Press, 2003.

Phelps family diaries and letters, personal collection, material also from archives in the St. Catharines Museum St. Catharines, Ontario.

Phelps, Oliver Seymour. *The Phelps Family of America*. 1870. Reprint: Pittsfield, MA: Eagle, 1899.

Phelps, Oliver Seymour (Junius). St. Catharines A–Z, 1856. St. Catharines Historical Society reprint. This is a collection of newspaper columns the effect of which is to demonstrate the fierce egalitarianism of many early settler families.

Pipes, Richard. *Property and Freedom: The Story of How Through the Centuries Private Ownership Has Promoted Liberty and the Rule of Law*. New York: Knopf, 2000.

Ridley, Matt. "Controlling the British Countryside." In *A Countryside for All,* edited by Matt Sissons, 2002, but here accessed at PERC Reports, vol. 20, Spring 2002, http://www.perc.org/articles/article260.php.

Scull, John. "Land Trusts and Conservancies in B.C.: A Survey by the Land Trust Alliance of B.C." Land Trust Alliance of British Columbia, 2003; also material from the Conservation Data Center; Land Trust Alliance of B.C.; Friends of Ecological Reserves, B.C.

Chapter Four

American Planning Association, Communications Boot Camp Webinar #1: First Steps: Responding and Reframing Planning. November 4, 1011.

American Planning Association. "Responding to Critics and Advocating for Planning." Heard on the Hill. Virginia Chapter Annual Conference, July 21, 2011.

Executive Order No. 13575, June 9, 2011. "Establishment of the White House Rural Council." *Federal Register* 76, no. 114 (June 14, 2011).

Howell, Peter, and Abigail Weinberg. "Mining Nature's Bounty: Tapping Markets for Ecosystem Services: Creating a Market Where There Is None." Open Space Institute, United States Business Council for Sustainable Development, 2005.

Lamb, Henry. "Sustainable Development or Sustainable Freedom?" *Range*, Fall 2010.

President's Council on Sustainable Development, Task Force Report. "Sustainable Communities," Fall 1997, http://clinton2.nara.gov/PCSD/TF_Reports/amer-intro.html.

Chapter Five

Abram, David. *The Spell of the Sensuous: Perception and Language in a More-Than-Human World.* New York: Vintage Books, 1997.

Arendt, Randall, and Elizabeth Brownbec. *Rural by Design: Maintaining Small Town Character.* Environmental Law Foundation, Lincoln Institute of Land Policy, American Planning Association, Planners Press, May 1, 1994.

Budiansky, Stephen. *Nature's Keepers, the New Science of Nature Management.* New York: Free Press, 1995.

Dixon-Hunt, John, and Peter Willis. *The Genius of the Place: The English Landscape Garden, 1620–1820.* Cambridge, MA: MIT Press, 1988.

Hiss, Tony. *The Experience of Place.* New York: Knopf, 1990.

Islands Trust, Victoria, B.C.

Kunstler, James Howard. *The Geography of Nowhere: The Rise and Decline of America's Man-made Landscape.* New York: Free Press, 1994.

McPhee, John. *Encounters with the Archdruid: Narratives about a Conservationist and Three of His Natural Enemies.* New York: Farrar, Straus & Giroux, 1971.

Pierson, Melissa Holbrook. *The Place You Love Is Gone: Progress Hits Home.* New York: W. W. Norton, 2006.

Regional Conservation Plan 2005–2010, Islands Trust Fund, Victoria, B.C.

Regional Conservation Plan 2011–2015, Draft. Islands Trust Fund, Victoria, B.C.

Sensitive Ecosystems Inventory, East Vancouver Island and Gulf Islands, 1993–97.

Steiner, Frederick R., and Robert D. Yaro. "A New National Landscape Agenda: The Omnibus Public Land Management Act of 2009 Is Just a Beginning." *Landscape Architecture*, June 2009.

Wells, Jeffrey, Dina Roberts, Peter Lee, Ryan Cheng, and Marcel Darveau. "A Forest of Blue, Canada's Boreal Forest, the World's Waterkeeper." Pew Environment Group, 2011.

Chapter Six

Anderson, Terry, ed. *You Have to Admit It's Getting Better: From Economic Prosperity to Environmental Quality*. Stanford, CA: Hoover Institution Press, 2004.

Anderson, Terry, and Laura E. Huggins. *Property Rights: A Practical Guide to Freedom and Prosperity*. Stanford, CA: Hoover Institution Press, 2003.

Brubaker, Elizabeth. *Property Rights in Defense of Nature*. Toronto: Earthscan Limited, Enviroprobe, 1996.

Emmett, Ross B. "Malthus Reconsidered: Population, Natural Resources, and Markets." PERC Policy Series, 2004.

Fleck, Robert K., and F. Andrew Hanssen. "Do Profits Promote Pollution? The Myth of the Environmental Race to the Bottom." PERC Policy Series, August 2007.

Goodman, Doug, and Daniel McCool, eds. *Contested Landscape: The Politics of Wilderness in Utah and the West*. Salt Lake City: University of Utah Press, 1999.

Hawken, Paul, Amory Lovins, and L. Hunter Lovins. *Natural Capitalism: Creating the Next Industrial Revolution*. Boston: Little, Brown, 1999.

Meiners, Roger E., and Lea-Rachel Kosnik. "Restoring Harmony in the Klamath Basin." PERC Policy Series, January 2003.

Morriss, Andrew P., William T. Bogart, Andrew Dorchak, and Roger E. Meiners. "7 Myths About Green Jobs." PERC Policy Series, 2009.

Simpson, David. "Conserving Biodiversity Through Markets: A Better Approach." PERC Policy Series, 2004.

U.S. Energy Information Administration. "Federal Financial Interventions and Subsidies in Energy Markets." Washington, DC: U.S. Department of Energy, 2007.

Chapter Seven

Arnold, Ron. *Freezing in the Dark: Money, Power, Politics and the Vast Left Wing Conspiracy*. Bellevue, WA: Merril Press, 2007.

Arnold, Ron. *Undue Influence: Wealthy Foundations, Grant-Driven Environmental Groups, and Zealous Bureaucrats That Control Your Future*. Bellevue, WA: Merril Press, 1999.

Arnold, Ron, and Alan Gottlieb. *Trashing the Economy: How Runaway Environmentalism Is Wrecking America*. Bellevue, WA: Free Enterprise Press, 1994.

Bacigalupi, Debbie. "Water Wars for the 21st Century: Evidence Based Management or Agenda?" Master's Thesis, Notre Dame de Namur University, School of Business and Management, December 2011.

McKenzie, Kirk. DefendRuralAmerica.com. Extensive interviews on dam removal effects, and evidence-based science on coho and dam removal, archived here.

Oakley, Marti, and Barb Peterson. "Vertical Farming, Forestry, Farming, Ranching." TS Radio, blogtalkradio.com, air date November 16, 2011.

Proposed Klamath River Basin Restoration Agreement for the Sustainability of Public and Trust Resources and Affected Communities. January 15, 2008, Draft 11, http://www.edsheets.com/Klamath/Proposed Agreement.pdf.

"Siskiyou Crest National Monument, America's First Climate Refuge." Klamath-Siskiyou Wildlands Center, January 2010.

Timber Harvest Levels on the Major National Forests in Siskiyou County. California Forestry Association, 1978–1993.

Vincent, Carol Hardy. "National Monument Issues." Congressional Research Service, Report for Congress, February 2005.

Chapter Eight

Two dozen number-rich reports detailing the catastrophe that fell on the men and women of Del Norte are archived on my website, and can be read in full. Del Norte County is a case study for the ages. Extensive charting of the money received and spent by the top twenty international ENGOs, foundations, and governments will be found there. Largely sourced from ENGO and foundation IRS 990s, this documentation provides an illuminating look at the financing of a top-down, masterfully planned assault on the public interest.

Chapter Nine

Arnold, Ron. "Battered Communities: How Wealthy Private Foundations, Grant-Driven Environmental Groups, and Activist Federal Employees Combine to Systematically Cripple Rural Economies." Bellevue, WA: Center for the Defense of Free Enterprise, 2000.

Capp, Diana White Horse. *Brother Against Brother: America's New War Over Land Rights.* Bellevue, WA: Free Enterprise Press, 2002.

Growth Management Act and Related Laws. Washington State Department of Commerce, 2010 RCW Update.

Chapter Ten

"Endangered and Threatened Wildlife and Plants: Review of Native Species That Are Candidates for Listing as Endangered or Threatened; Annual Notice of Findings on Resubmitted Petitions; Annual Description of Progress on Listing Auctions." U.S. Department of the Interior, Fish and Wildlife Service. *Federal Register* 76, no. 207 (October 26, 2011).

Extensive citations—sacred and profane—on supposed biodiversity loss, the endangered species act, and the litigation it has spawned, many with hyperlinks, can be found on my website.

Karen Budd-Falen's research on litigating green groups and their looting of the public purse may be found on the website of the Western Legacy Alliance, or on Budd-Falen's law office website: http://www.buddfalen.com/eaja.html.

Chapter Eleven

Garrison, Trey. "Not One More Acre! Ranchers in Colorado's Pinon Canyon Fight a Massive Army Land Grab." *Reason*, February 17, 2009.

Gattuso, Dana Joel. "Conservation Easements: The Good, the Bad, and the Ugly." National Policy Analysis, National Center for Public Policy Research, #569, May 2008.

Lewis, Kimmi Clark. "Conservation Easements, the Colorado Experience." Powerpoint presentation, archived on www.elizabethnickson.com.

Merenlender, A. M., L. Huntsinger, G. Guthy, and S. K. Fairfax. "Land Trusts and Conservation Easements: Who Is Conserving What for Whom?" *Conservation Biology* 18, no. 1 (February 2004).

Parker, Dominic P. "Conservation Easements: A Closer Look at Federal Tax Policy." PERC Policy Series, October 2005.

Parker, Dominic P. "Maintaining Trust: The Challenge Facing Private Land Conservation." PERC Reports 21, no. 2, June 2003.

"Working with the Nature Conservancy." Memorandum of understanding between the U.S. Department of the Army, the U.S. Army Corps of Engineers, and the Nature Conservancy, December 2000, archived on www.elizabethnickson.com.

Chapter Thirteen

American Stewards of Liberty, www.americanstewards.us, holds a wealth of information on the statutes that require coordination and describes the process by which local governments can force federal and state agencies to the table to reach consistency with local economic needs, custom, and heritage.

Chapter Fourteen

Additional Chuck Leaf research, with charting, is archived at www.elizabethnickson.com.

Leaf, Charles F. "Cumulative Hydrologic Impacts of U.S. Forest Service Management Practices on the Routt, Arapaho-Roosevelt, Pike, and Medicine Bow National Forests: Potential for Water Yield Improvement." Platte River Hydrologic Research Center, 1999.

Leaf, Charles F., and Forrest A. Leaf. "Hydrology and Well Augmentation in the South Platte River Basin." Research Paper PRHC–9, Platte River Hydrologic Research Center, July 2008.

Meiman, James R. "The Legacy of Collaborative Watershed Research Between the Rocky Mountain Forest and Range Experiment Station and Colorado State University." Hydrology Days, Colorado State University, 2005, http://hydrologydays.colostate.edu/Papers_2005/Meiman_paper.pdf.

Chapter Fifteen

Keeler, Judy. "The Wildlands Project Comes to Hidalgo County." Keeler's sixteen-part series with sourcing is archived at http://www.uhuh.com/1calfraud/stacks/judymusin.htm.

INDEX

ABOUT THE AUTHOR

Columnist, investigative journalist, and novelist Elizabeth Nickson has been a national columnist for Canada's *Globe and Mail* and *National Post*. She was European bureau chief of *Life* magazine and a reporter for *Time* magazine, and has written for many international publications, including the *Sunday Times Magazine* (London), the *Guardian*, *Tatler*, *Vogue*, and *Harper's* magazine. Her novel *The Monkey Puzzle Tree* examined the CIA mind control research program in Montreal in the 1960s. She lives on Salt Spring Island in the Pacific Northwest.